Benjamin Ward Richardson

Public School Temperance

Lessons on Alcohol, and it's Action on the Body

Benjamin Ward Richardson

Public School Temperance
Lessons on Alcohol, and it's Action on the Body

ISBN/EAN: 9783744670531

Printed in Europe, USA, Canada, Australia, Japan

Cover: Foto ©berggeist007 / pixelio.de

More available books at **www.hansebooks.com**

PUBLIC SCHOOL

TEMPERAN

LESSONS ON ALCOHOL, AN
ACTION ON THE BODY.

DESIGNED FOR PUBLIC SCHOOLS.

BY

BENJAMIN WARD RICHARDSON,

M.A., M.D., LL.D., F.R.S.,

FELLOW OF THE ROYAL COLLEGE OF PHYSICIANS, HONORARY PHYSICIAN TO THE ROYAL
LITERARY FUND, AND AUTHOR OF THE CANTOR LECTURES ON ALCOHOL.

TORONTO:
THE GRIP PRINTING AND PUBLISHING COMPANY.
1887.

PREFACE TO CANADIAN EDITION.

In compliance with a well understood public opinion an Act to provide for the teaching of Temperance in the Public Schools was introduced at the last session of the Legislature of Ontario, and received the unanimous approval of the House. Under the provisions of that Act, the subject is placed in the Programme of Public School studies ; and this volume is authorized by the Department of Education as the text-book.

The author of this work is the celebrated Dr. Richardson, F.R.S., F.R.C.P., of England, Hon. Sec. to the Royal Literary Fund, and author of the Cantor Lectures on Alcohol, who is known throughout the British Empire as one of the highest authorities on this and kindred scientific subjects. The publishers have secured the copyright for the Dominion of Canada, and, under the direction of the Department, have slightly re-arranged the matter so as to adapt it more fully to the capacity of Public School pupils.

The book will be sold to pupils and the public for twenty-five cents. It will not fail to be observed that, though containing all of the matter of the English edition, it is only one-half of the price.

The plan of the work will readily commend itself. Each branch of the subject is presented in simple language, and arranged in a short lesson, to which questions are appended. The information imparted, while of the most interesting and valuable character, is made as general as is consistent with a knowledge of the subject ; the intention being not so much to give details and technicalities, as to explain the effects of alcohol on the human system, and to impress the pupil with the danger of its use.

iii.

The fact that many contract the habit of using intoxicating liquors through ignorance, and that even the best education imparted in our Public Schools is not an offset to the ruin which frequently results, may be accepted as ample justification for the course which our legislators have now taken. What is learned in childhood usually exerts an influence for life ; and it is believed that this new subject will not only prove an interesting and valuable addition to our Public School course, but will have an important moral effect on the lives of the coming men and women of our country.

CONTENTS.

PUBLIC SCHOOL
TEMPERANCE.

LESSON I.

ARTIFICIAL DRINKS.

For ages men and women of many countries have learned to drink what are now called spirituous liquors. These liquors have been taken in various forms, and various names have been applied to them. In our times the commonest names are ale, beer, stout, porter, gin, rum, whiskey, brandy, and wine.

Ale, beer, stout, and porter are usually called malt or malted liquors. Gin, rum, whiskey, and brandy are called spirits, and wines of all kinds are spoken of under the general name of wine. A person who indulges particularly in the use of these fluids is often named by the connection. He is said to be a beer-drinker, a spirit-drinker, or a wine-drinker.

Some people think when they are speaking of beer, spirit, or wine, that they are speaking of things which are quite distinct, and which act in a distinct manner when they are taken into the body as drinks.

In like manner some people think that wines bearing different names are distinct : that port is distinct from sherry, sherry from Madeira, Madeira from claret, and so on, and in some degree this is true, because the colors of these various fluids, as well as their odors and their tastes, are very distinct one from the other.

In our country there are in common use ales and porters

of several kinds, gins, and whiskies, and rums, and brandies of several kinds, and I have counted as many as fifty various kinds of wines coming from other countries, to say nothing of wines which are made at home, and to which such names as cowslip, ginger, raisin, and gooseberry wine are applied. Sometimes we hear of other drinks, as of cider or perry, and of sweet and very strong drinks called liqueurs. These in respect to appearance, smell, and taste, differ from each other, as well as from beers and wines and spirits.

To persons who have never tasted these drinks many of them are nauseous when first tasted. Even to grown-up men, who have never before taken these liquids into their mouths, the first taste is like that which is felt on taking a medicine. The taste is said to be bitter in respect to the ales, clammy and sickening in respect to porter and stout, burning and sickening in respect to spirits, and burning and sour, or burning and sweet, in respect to the wines.

In all my experience I never once knew a person who liked the first taste of any one of the drinks we are now thinking about. This fact seems to me to show clearly that it was never intended that human beings should take these drinks regularly every day. If that had been intended, the drinks would have been made and given to us in a form that would have been pleasant to the taste, or at all events in a form that would not be so unpleasant, that the instinctive or natural feeling is opposed to them. Water and milk are natural drinks. They are neither bitter nor nauseous, nor acid, nor burning, and therefore even the youngest infants and children take them without dislike, and look for them with quite a longing desire when they want drink.

It is a lesson early to be remembered, and so I write it down early in this book, that although there are so many drinks made and sold as beers, wines, and spirits, none of them are fitted to the first natural wants and desires of man. I gather from the facts before us that the said drinks are therefore not wanted at all. If a little child can live and grow up, and learn and work and play, and be very healthy and pretty, and strong and happy, without these drinks, a man and a woman can live without them equally well. We call all these strong drinks artificial, in order to distinguish them from the natural drinks, water and milk.

Questions on Lesson I.

1. What are the names of the various liquids, called generally spirituous drinks?

2. When these drinks are tasted for the first time, what are the feelings they usually excite?

3. How do water and milk differ from them, as drinks, in regard to taste?

4. What do you mean by instinctive tastes?

5. What do the facts respecting the instinctive tastes in regard to strong drinks teach?

6. What do you call strong drinks, in order to distinguish them from natural drinks, like water and milk?

LESSON II.

NATURAL DRINKS.

WE found at the close of our last lesson that persons who have never tasted artificial strong drinks may grow up perfectly well without them, and be very healthy and very happy. It is right to remember that this same rule applies to all the lower animals, and that, too, through every period of their lives. The lower animals crave for one fluid alone as drink, viz. for water. Many of them, all that are called mammals, or milk-giving animals, live, when they are young, on the milk which their dams yield to them, but after they are weaned they wish only for water; while many other animals that are not mammals, have never at any time any drink except water. It is important to think what a wonderful amount of life and of work is carried on, and what varied work is done, on simple water, by other animals than man. With what speed some birds can fly, some horses gallop: to what heights some animals, such as the deer, can leap: what weights some animals, such as the elephant, can carry: what amount of exercise a dog can go through, and what intelligent tasks it can be taught to perform. The fleet camel can walk and carry its rider fifty miles in the day and can exist in the arid desert for days, supplied with the water which it carries in great water-cells or pouches in its body. All the animal kingdom, indeed, mam-

mals, birds, fishes, reptiles, insects, microscopical animals, or
those which only become visible when they are looked at
through a powerful magnifying glass—all this vast world of
life, extending from the Polar seas to the heated tropics, in
the sea, in the air; all these myriads of forms of living motion
can go through their destined tasks, their sports and works,
having as their drink nothing but water. To think of these
facts is to feel the best of proofs that artificial fluids have no
place whatever in the scheme of creation, and that the natural
human instinct which, as we have seen, abhors at first artificial
drinks, does so because it would, if it had its way, lead men
through the same simple process of living as it does the less-
endowed animals which have acquired no artificial or inventive
skill whereby to manufacture substances that the earth does
not supply directly from her own store.

We understand this matter still better when we study the
composition of living bodies, whether the bodies of men or of
the inferior creation. Into the building up of these bodies
water enters so largely that some forms of life are almost
entirely formed of it. In the sea by the coast there are often
found animals of a jelly-like kind which are called by the
people "jelly-fishes," and by the learned, medusæ. One of
these, which I collected while in its fresh state, weighed one
hundred and forty ounces; but when it had lost its water by
gentle drying, all the solid matter which it contained weighed
no more than the eighth part of an ounce, so that this animal
was actually made up of one thousand one hundred and nine-
teen parts of water to one part of solid substance. This is
an exceptional instance, adduced merely to show what a
wonderful part water plays in living bodies. It plays also a
very remarkable part in the human body. The blood contains
seventy-nine parts of water out of the hundred; those engines
of the body, the muscles, which move the limbs and other
parts, contain seventy-five parts of water in the hundred; the
brain, which is the organ of the mind, contains no less than
eighty parts of water in the hundred.

We cannot, without a knowledge of facts like these, understand the true value of water to man and other living crea-
tures. It is the great essential to life; but that it may be
essential, it must exist as water and nothing else. If any other
fluid be mixed with it in the body, and by the mixing, alter its

weight or other properties, it ceases to perform its duties correctly, and if in the blood of any young animal there were found any other fluid, we should say that the animal was charged with a substance that did not naturally belong to it.

It may perhaps occur to you who are reading this lesson that in the bodies of animals there are fluids which are known by other names than water. You will say there is saliva and gall or bile, and other fluids, and that sometimes animals yield milk. This is true; but all these fluids are made up of water, and are only different from water because they contain solid substances dissolved or suspended in the water, so that water is the fluid basis of every one of them.

Questions on Lesson II.

1. On what simple fluid do all animals as well as man naturally subsist?

2. Give some illustrations of what kinds of work and exercise animals can do on water alone as drink?

3. What animal is endowed to carry water in pouches within its own body over long journeys?

4. What quantity of water exists in the human blood?

5. What quantity of water is found in the muscles, and what in the brain of man?

6. Is any other fluid found naturally in the bodies of man or other living animals, and why do other fluids of the body seem to differ from water?

LESSON III.

THE WATER OF THE BODY.

THE water in the bodies of men and of all animals plays many varied parts. No other known fluid could in any way perform the same important duties. I have already said that water is the principal substance of the vital organs of the higher animals, such as the brain, the muscles, the blood; and I showed that in the jelly-fish it formed nearly all the body. I may add in this place a note of an observation I once made on a human body in respect to this same subject. I took one

of those remains of the human body which have been preserved
some thousands of years, and which is called an Egyptian
mummy.   It was the body, probably, of one who had been a
great priest or ruler, for it had been embalmed or preserved
in the most expensive form of embalming, and had been enclosed in a sarcophagus or tomb, which must have cost a little
fortune.   I measured the mummy, its length, its girth, and
the relative size of its head and limbs and trunk.   From these
measurements I was able to estimate what would have been
the weight of the body when its owner was moving on the
earth in the midst of life and health.   The weight of the body
at that time, 1 reckoned, would have been one hundred and
twenty-eight pounds.   In the condition of a mummy in which
it was now before me, nothing remained but the dried skeleton
or bony framework, and muscles and other organs completely
dried.   The body in fact had, in course of ages, lost all its
water.   In this state it weighed just sixteen pounds, and as
eight sixteens are one hundred and twenty-eight, it is clear
that seven parts out of eight of the whole body, or one
hundred and twelve pounds, had passed away as water.   In
the remaining weight was included that of the skeleton, which
contains but ten per cent. of water, and some mere remnants
of canvas and pitchy substances, which had been used by the
embalmers, and which, like the skeleton, continued still perfect.   The soft parts of this human body, by which all its
active life, its moving and thinking functions, had been
carried out, were in fact nearly all removed by the drying process, or loss of water, to which they had been subjected.   They
were not decomposed, they had not, that is to say, been de-
stroyed by passing into new forms of matter, as occurs when a
dead substance is allowed to decay in the open air, but they
had completely lost the water which once gave to them size,
flexibility, shape and capacity for motion.   Water, then, in
the living body, gives to the parts which move and are active
their chief weight and size, and all the soft tissues of the
body may be looked upon as water rendered partly solid.
How this solidity is brought about it is well to know.

Some organic substances, by which I mean substances
which are found in animal bodies only, have a peculiar affinity
for water, by which they are able to take up a great quantity
of water into themselves, and so assume a soft or jelly-like

appearance. Common glue is a substance of this kind. If glue be dried it can be brought into such a state of dryness that it will break almost like glass, but if it be exposed to water it will take up large quantities, increase in weight, and when the water to which it is exposed is hot, it will pass into a liquid. Jelly will do the same, and these substances are taken as types of many more, and are called *colloids*. Muscles, membranes, the skin, two portions of the blood known as fibrine and albumen, a part of the brain and of the nerves, not to name many other parts of the body, are colloidal parts. Through them water is diffused; they are *hydrated*, that is to say, charged with water; and so equally is the water distributed or spread through them, that in some instances they are quite liquid, like water itself. The blood flowing through the blood-vessels of the body is, for example, a liquid. It is by being diffused in this way through the colloidal or jelly-like organic substances of the body that water so largely exists in the body. It is from these same substances it escapes when it is removed by the process of drying. There is no other known fluid that will in like manner diffuse equally through the colloidal substances of the body, to render them plastic, or yielding, and, when necessary, liquid. Some other fluids will dissolve colloids, as ether does when it dissolves gun cotton (which is a colloid) and makes collodion; but nothing, except water, acts in the same way on the colloids of the living bodies of men and animals.

Questions on Lesson III.

1. How much water is given up by the human body from the process of complete drying?

2. From what parts of the animal body is the water that escapes in drying given off?

3. What is meant by an organic part or structure of an animal body?

4. What is meant by the terms colloid, and colloidal substance?

5. What does water do to the colloidal parts of the living body?

6. Will any other known fluid, besides water, act in the same manner on the colloids of the body?

LESSON IV.

USES OF WATER IN THE BODY.

WATER not only gives form, size, flexibility, and capacity for motion to the colloidal parts and organs of the body, but it performs many other equally useful purposes. For instance, it is the great solvent of the different substances which we take in as food, and of the juices by which food is digested. When we take solid food into our mouths, and chew or masticate it, it becomes mixed with a fluid in the mouth called the saliva, which fluid contains ninety-nine parts out of the hundred of water. Here the food is softened and a small part of it dissolved. When the food is swallowed and enters the stomach, it is mixed in the stomach with another fluid called the gastric or digestive juice, which fluid contains ninety-seven parts out of the hundred of water. Here the parts of the food that are made up of colloids, the fleshy or muscular part of animals on which we feed, the substance of eggs, the cheese that is in milk, the gluten that is in bread, are dissolved and made ready to enter and to circulate through the body in the water of the blood. After the colloidal food is received and digested in the stomach, there are other foods of a starchy or fatty nature which pass out of the stomach undigested, with much of the food that has been digested, into a further part of the digestive canal. Here these as yet undigested foods meet with two other fluids called respectively the bile and the pancreatic juice. The bile contains eighty-seven parts of water in the hundred parts, and the pancreatic juice ninety parts in the hundred. By the action of these juices the food which passed through the stomach unchanged is brought into solution and prepared to enter the blood.

In this way all our food is brought into solution or suspension in the water of the various digestive juices I have named, and in that which was swallowed with it, and so it is fitted to be absorbed or drawn into the blood-vessels, and into the circuit of the blood.

14

But for the agency of water not one of the changes we have seen could have been carried out, and but for the water within the blood, the food could not afterwards be carried as it is, as by a river, to all portions of the body to build up or renew those structures which are constantly being removed by the work of the muscles and other active organs, and which must be re-supplied in order to prevent death from wasting and exhaustion.

Besides the colloidal and starchy and fatty foods which the water helps to bring into solution, there are other substances which we take in by or with foods, and which are called salts. Common table salt, *chloride of sodium*, is a good example of this nature. These salts are very soluble in water, and the water is the means of conveying them in solution into the blood. The blood contains common salt, phosphate of soda, and some other salts, which give to it its peculiar saltish taste, and all of these are carried in the water of the blood around the system. The salts add to the weight of the blood, they help to keep the colloids in equable solution, and they form the hard part of the solid skeleton.

We learn from all these facts, then, that water is the grand solvent of all the parts of the body, and that every part, however firm and solid it may seem to be, has once been in a state of solution in water before it became solid. This is true even of the bones which form the skeleton. They contain twenty-three parts of organic matter in the hundred, and they contain sixty-seven parts of inorganic matter in the hundred, the constituents of which were carried by water. They also contain ten parts in the hundred of water itself, so that bone, hard as it is, is brought into form by water, and this is equally true of all the other organs, that their parts come together by the agency of water. No other fluid that we have any knowledge of could have performed the same duties except water, and it is so nicely and delicately adapted for its duties that if any other fluid be introduced into it it is impaired in its usefulness.

By some means, which are not as yet fully understood, the force or power by which the animal body is moved is brought into action by water. Some animals which have a very simply constructed form of body, may be deprived of their water during long exposure to heat in tropical climes, and may seem to be actually dead and dried up. But on being exposed to water they will revive, regain their shape, and once more live and move.

Questions on Lesson IV.

1. To what other substances does water act as a solvent?

2. What proportion of water is there in the gastric or digestive juice?

3. What proportion of water is there in the bile, and in the pancreatic juice?

4. What parts of food are brought into suspension or solution by water?

5. Where does the water convey the food after its solution and its entrance into the blood?

6. Would any other known fluid play the part of water in these processes of digestion and conveyance of food?

LESSON V.

THE WATER-CURRENT IN THE BODY.

WE have not yet finished with the uses of water in relation to the living body. We have seen that it forms a great part of the structure of the body, and that it is the fluid in which the food we take is digested and dissolved in order to be carried into the blood, and to replace so much of the animal structure as is lost by wear and tear. There are, however, yet other uses performed by this simple but remarkable agent, water, and to these uses we may, with profit, direct our thoughts.

When the blood, which contains seventy-nine per cent. of water, passes through the blood-vessels called arteries, into the extreme parts of the system, it carries with it in solution in water the various substances out of which the different organs of the body are composed. These are separated out in the different organs by means of the water. A series of fine membranous screens interpose between the blood and the tissues, and through these membranes the colloids will not pass. But the water, in due proportion, will pass through the membranes, and the salts will pass with the water, fixed in fact with it. In these refined portions of the living animal body great and important chemical changes take place. The used-up or worn parts of the organs are changed, and are conveyed away, in the soluble saline form, by the current of water. In the same parts animal warmth is produced by the process of chemical change that is in progress, and this warmth is, I believe, distributed by the water. In various parts of the body there are placed little spongy structures called glands, in which water is separated from the blood, together with the salts which have been produced by the breaking up of the used-up food, and which water and salts, by means of ducts or tubes springing from the glands, are removed from the body, that the body may be purified of all that is useless.

Thus water not only carries into the body those things which are requisite for subsistence, but carries out also those things which have served their purpose and are therefore unnecessary. The fluid which escapes from the skin, and which is known as the sweat or perspiration, is water that is being carried away, bearing other products which have served their purposes in the economy.

It will be supposed from this description that there must be a constant current of water through the body during life, a current which enters through the stomach by being swallowed as drink, which goes into the blood, carrying with it the food, and which escapes by the outer surfaces of the body, carrying with it the *débris* or used-up food. The supposition is quite correct. This is precisely what is always going on so long as we live in health, and if it is not going on correctly we are not in health. If the quantity of water taken into the body be not sufficient, there is disturbed balance, fever, and, in time, death. If the water, which should pass out of the body by the external surfaces, be stopped from passing out, there is disturbed balance, coldness of the body, accumulation of water in the organs and cavities of the body (dropsy), sleepiness, and, in time, death. If the water be drawn off too rapidly, there is disturbed balance, coldness, shrinking, convulsions, or spasms of the muscles, and, in time, death. If, again, the water within the body be changed in its quality, as by the addition to it of an excess of waste matter of the system, there is disturbed balance, which may end in convulsions and death.

Questions on Lesson V.

1. What fine structures interpose between the blood and the solid parts of the organs of the body?

2. What happens to the colloids in reference to these structures?

3. What happens to the water, and to the salts the water contains?

4. What changes take place in those organs of the body which are called glands?

5. What does water do besides the duty of carrying food into the tissues of the body?

6. Describe the course of the current of water that is in progress through the body during life.

LESSON VI.

WATER-DRINKERS.

In the Divine scheme and order of Nature there is profusely provided one fluid for the wants of man and of other animals as drink. Living bodies are built for water only to be the fluid by which they may live, and up to the present time this plan has not been disturbed by man in respect to any animal inferior to himself. Domestic animals, (horses, sheep, oxen, dogs, cats,) wild animals, (lions, leopards, tigers, bears,) large and powerful animals, (camels and elephants,) these have all continued to live on water as their simple drink. Even those animals which are nearest to man in their appearance and anatomy, (monkeys and apes,) even these have always continued to exist on water. In fact, throughout all the range of the animal world, every animal, large and small, strong and feeble, fleet and slow, has found water the all-sufficient drink. It is man alone who, by the aid of his unfettered reason and skill, has changed the natural ordinance, and used as a drink a fluid different from that which he received spontaneously from the bounteous Hand by which he was already amply and correctly provided.

When I say it is man alone who has ventured to change for himself the provisions of Nature, I do not wish to include in that saying all men, men of every race, of every country, of every age. If we could collect the history of all the times and peoples of the earth, we should, I believe, find that after all only a small part of those men and women who have lived onthe earth up to this time have broken the natural law. The great multitudes who have lived in the past have lived on water as drink, and at this very day there are millions of people who find water all-sufficient for their necessities.

Moreover, amongst those peoples who have invented different drinks, there have been none who have not mainly subsisted on water. No one has yet lived, actually lived, on any other fluid, because no one could live on anything else as drink. The worst persons who break the natural law retain water as their staple drink, though many are so blindly ignorant they do not know the fact. They drink what they think is not water at all, and they give to the drinks so used many very fine names. They pay for the fluids they drink very heavy prices, they go to a great deal of trouble in order to obtain them. But strangest of all is that, in spite of fine names, expense and trouble, they only succeed at last in getting a drink the larger part of which is water. They get a drink which would not be drink if it were not so largely made up of water, and which is indeed water deprived of its purity, water to which something has been added that interferes with the proper action, but which is water in whatever disguise of name, or color, or taste it may be presented.

So very ignorant are many people on this subject that they assume to laugh at what they call "water-drinkers." They themselves declare they never drink water, and perhaps they profess to pity those who do. They hold up some sparkling or colored liquor in the glass and boast, "This is my drink ! None of your slip-slop water for me !" Poor people ! why, at this very time they are holding up a drink which has in it at least three parts out of four of water, as we will see in a lesson which soon will come. They are drinking slip-slop water also ; they have merely added another slip-slop which has no business there, and which spoils the natural intention. If they had not so much water, few of them could get the fluid down their throats, so strong is the opposition of Nature to any foreign liquid. Water, therefore, in these cases is positively necessary to make that which is undrinkable drinkable at all, so that it is both false and foolish for any one to boast that he is not a water-drinker.

* * *

Questions on Lesson VI.

1. What general evidence is there that water is an all-sufficient drink?

2. What special evidence is there from man that water is an all-sufficient drink?

3. Is there any other fluid than water on which men and animals can exist?

4. What is a very common mistake respecting other so-called drinks than water?

5. What do persons who indulge in other fluids subsist on, after all, as their drink?

6. Is there any drinkable drink without water?

* * *

LESSON VII.

NATURAL FOOD.

As water plays such an important part in the bodies of living animals and of men, we may be sure it is necessary at all times that it should be taken into the body in proper quantity and in due proportion to the other articles of food that join it for the support of life. It would be good, therefore, if we could find a standard by which to determine what proportions Nature herself fixes. Fortunately we have such a standard in the food which Nature provides for so many animals at the period of their lives when they are too young to take care of themselves and to procure food by their own efforts. The food which gives us this standard is called milk.

We shall do well to learn the composition of this standard food, for it not only is capable of supporting the life of the very young, but it is also capable of supporting the life of those who are advanced in years. Sometimes when persons are ill they are fed on milk solely, and I know one person, at least, who for more than twenty years has never taken any other food, and who lives on milk, pure and simple, with excellent health and strength.

Milk, then, is composed of a solid and a fluid part. The solid weighs twelve parts in a hundred, the fluid eighty-eight. That is to say, if one hundred ounces of new milk be heated carefully, so as to get rid of all the fluid portion, the quantity left of solid food will be twelve ounces; and if all the fluid that was carried off in the process of drying be collected, as it could be by condensing it, it will be found to weigh eighty-eight ounces. The solid part of the milk, when that is analysed, is found to consist of a colloidal food called caseine—cheese; of a fatty substance—butter; of a sweetish substance—milk sugar; and of salts. These solid parts are nicely divided so as to meet all the requirements for building up new tissues, for maintaining animal warmth, and for supplying the due amount of salts. But that which concerns us most is the fluid part; the eighty-eight ounces of the fluid which holds all the other portions of the milk either in solution or in suspension,—of what is it composed? It is simply water.

And so we know by this standard food, supplied by Nature herself for the use of so many animals, that water is the only drink with which she mixes her solid food, the only drink she, in her wisdom and beneficence, provides for them.

With the first food of man Nature gives water. In various other modes she also gives the same. There is no palatable solid food in which some water is not mixed by her hand. Bread contains water, animal flesh contains water; all the fresh products of the vegetable world, fruits, and herbs, and grasses, contain water, many of them in abundant quantity. Some animals, such as the rabbit, find in their vegetable food so much of water that they can live naturally without drinking from any stream or fountain. In addition to these sources of liquid food, Nature, with the most liberal hand, distils for us the pure liquid. From the earth the water rises, in distillation, under the heat of the sun. Ascending in vapor it becomes pure in the air. It condenses on the mountain-tops from the vaporous into the liquid form, and streams down in rivulets and torrents. It condenses in clouds and falls in showers of rain. It is stored up for us in cool reservoirs of the earth, or runs to us in rivers and brooks. It answers to our early cravings of thirst, and it would answer to our last, without any substitute for it, if we would but let it.

Questions on Lesson VII.

1. What is the best example of a natural standard food?

2. What is the proportion of solid and fluid matter in this standard food?

3. What are the divisions of the solid parts of the food, and the uses of each part?

4. What is the fluid part of the natural standard food?

5. In what other foods does Nature supply the same fluid?

6. By what other modes does Nature supply the same fluid abundantly?

LESSON VIII.

WINE AND STRONG DRINK.

IN early ages of the world man, still a child in learning, looking like a child after his own delights and pleasures, and not foreseeing how many childish pleasures would seem unfitting in maturer and riper ages of thought, sought to gratify his tastes with other fluids, as he fancied, than simple water. One of these he called wine. He derived wine from fruit, and it is clear that he thought it so different from water as to have no idea of the fact that his new liquor was, after all, nothing more than a fluid diluted with water six or seven times at least. He did not understand, that is to say, that wine as it is obtained from the grape by the process of wine-making does not, when it is strongest, contain more than one part of the new fluid that makes it wine to six or seven parts of water. He therefore thought it, judging of it, naturally enough, by its difference of color, taste, and effect, a drink entirely new, a drink quite distinct from water. It was not until many ages had passed away, and the science now called chemistry had become a portion of human knowledge, that his mistake was found out and made known.

Where man first produced wine is a very difficult question to solve, and how he came to make it is one hardly less difficult. By some accident it was observed that if the juice of certain fruits, the juice of the grape especially, were exposed to the air, under certain circumstances, it began to change, to yield a froth, and to move as if it boiled. This appearance gave origin to the term "ferment," from the Latin verb *ferveo*, to boil. Then it was found that when the froth was skimmed away, a new and colored fluid was left underneath. In course of time some one, from necessity perhaps at first, or from curiosity, tasted this fluid, and finding that it had a very peculiar and exhilarating effect on him, gave it to others to taste or drink. In this way this fluid, which we call wine, became a luxury or treat on great occasions, for none at first held it to be a fluid that would take the place of water. From the mode in which it was used it is clear enough that wine was at first only intended to be placed among the luxuries. It formed part of the feast at times of great rejoicings and excitements. It was not long before it was found out to be very bad liquor for men who were about to perform great deeds and feats of prowess or skill, and it was also soon found to be a bad liquor even at the feast if it were partaken of too freely. Solomon, the writer of the Proverbs, is very decided on this point, and many ancient and modern writers, as we shall see by-and-by, were equally decisive in their opinions on the same subject.

I hope these important truths are now understood. First, that wine, which is water containing a new liquid, did not come down to man flowing as pure water flows from the earth, a liquid ready-made as drink for all animals, including man. Secondly, that the liquid which, added to water, makes wine, does not exist ready-made as water exists springing from the earth. Thirdly, that in order to obtain the fluid which so changes water as to transform it into wine, it is necessary to let the juices of fruit mix with water and undergo a change, called fermentation, and that if this process were not constantly being performed in the most wholesale way by men who devote their lives to the work, there would be no supply of wine or strong drink at all.

These facts will best be retained in the mind by recalling the terms that are applied to the processes by which wine and other strong drinks are produced. We talk of "home-made" wines, of "foreign-made" wines, of "brewing of beer," of "manufacturing gin and rum and other so-called raw spirits." By these terms it is meant that man produces the fluids by his ingenuity and labor.

Questions on Lesson VIII.

1. What was the first strong drink which man used?
2. How was this drink derived?
3. What great error did the men of ancient times hold in respect to this drink?
4. What process has the juice of fruit to go through in order to pass into wine?
5. On what occasions was wine first used by mankind?
6. By what simple terms do we express that wines and other strong drinks are manufactured for use?

LESSON IX.

WINE AND BEER IN ANCIENT TIMES.

FOR a great many ages wine was the only name given to strong drink. Why the name was first given is matter of great interest in history. The ancient Greeks, who were very proud of being connected with the introduction of wine into the list of substances used by men for drink, said that the drink was first discovered in Ætolia, a district of ancient Greece, and that the name wine comes from *oinos*, the vine, or from Œneus, a man who first pressed the ripe grapes, and from the juice made wine. The Romans called the fermented fluid *vinum*, and we, altering the word a little, call it *wine*.

When the ancients drank wine at their feasts they mixed it with water, and at first they used it very sparingly. It was offered at the sacrifices to the gods, and young men under thirty were not permitted to take it at all. Women were not permitted to partake of it at any age. In time these restrictions were broken through, and then wine began to be feasted on too liberally, to the injury of those who departed from the primitive rules or fashions.   This mistake has been commonly felt by those who have commenced to drink wines and other strong drinks.   At first they take them as luxuries only, in order, as it is said, to cheer up the spirits and make the heart less heavy.   In time the cheerfulness so obtained begets desire for repetition.   Men desire, naturally enough, to be cheerful at all times.   They say, if we could be made cheerful yesterday, why cannot we also be made cheerful to-day? and if to-day, why not to-morrow?   So the habit once started of seeking for cheerfulness from wine is continued until the habit passes into an artificial necessity.   In Roman times this occurred, until at last when the feasters sat down to wine, they were induced sometimes to have a skeleton brought in before them, on which he who gave the feast cast his eyes and bid his guests enjoy themselves while they could—a sentiment which in modern times has been expressed in what is called a jovial song :

> "Merrily fill up your cup to-day,
> To-morrow may find you a mould of clay."

And, alas ! it too often happens that the morrow does find those who drink wine moulds of clay before their natural time for death.

When men and women so desire wine that they must needs have recourse to it for daily happiness, they are brought already by the effects of the drink into close relationship with the skeleton from which the ancient reveller drew his false lesson.

Besides wine the ancients invented other fermented strong liquors like to wine, one of which inventions is still in common use under the name of beer.   The Egyptians are said to have first made beer by pouring hot water on barley and allowing the fluid to ferment.   It has been said that they called this drink "bouzy," from Busiris, the name of a city which contained the tomb of the god Busiris or Osiris.   So, says one of our quaint old authors, we get the term "bouzy," which we apply to a man who has taken a great deal of beer, and whom we call a bouzy fellow.   The word beer probably comes from barley, or from the Hebrew word *bar*, corn.

These were the origins of the strong drinks first invented,

and wines and beers of various kinds were introduced among men. Other kinds of similar drinks were discovered, but these need not now occupy our time.

Questions on Lesson IX.

1. Where was wine said to be first made, and from what is the word originally derived?

2. Under what conditions, and on what occasions, was wine first used?

3. What is the common result of taking wine on particular occasions?

4. What ceremony did the Romans sometimes perform at their feasts of wine?

5. What is the usual consequence when wine is resorted to regularly as the source of what is called happiness?

6. What other modern strong drink resembling wine in its effects was invented by the ancients, and whence did it derive its name?

LESSON X.

WISE MEN ON WINE.

I HAVE said already that many of the ancient wise men soon discovered that wines and strong drinks were very dangerous luxuries. It is well that some of the sayings of these wise men should be fixed in the minds of young people. Solomon, the wisest of men, knew perfectly well the bad effects of wine on those who indulge in it, and what he has said shows to us that in his time wine was a great source of evil. Let me copy down a few of his proverbs.

"Who hath woe? who hath sorrow? who hath contentions? who hath babbling? who hath wounds without cause? who hath redness of eyes?

"They that tarry long at the wine; they that go to seek mixed wine.

"Look not thou upon the wine when it is red, when it giv-eth his color in the cup, when it moveth itself aright.

"At the last it biteth like a serpent, and stingeth like an adder."

In these days when science has done so much to give us knowledge about wine and its action on the human body, it would be impossible to say more or use words more truthful than these ancient words. We know still from what we see in those who take much wine that it makes such persons full of woe when the excitement from it is over, that it makes them sorrowful and sad, that it makes them angry and quarrelsome, that it makes them talk foolishly about subjects they would be wiser not to mention, so that they do indeed babble ; and that it makes them think they have grievances and troubles which other men have not, when, in fact, none may exist.

Other men have said the same things as the wisest man said. The saying "In vino veritas" means, for example, that when men are full of wine they let out things about themselves which they would not let out if they were quite sensible. Again, the sentence "When the wine's in, the wit's out," tells the same truth.

But no one has ever told the truth so clearly as Solomon did in the proverbs I have quoted. It is very curious to observe that Solomon speaks of the appearance of the eyes of those who indulge in wine and strong drink. He says the eyes of these persons are red. It is the direct effect of wine to make the eyes, and not the eyes only, but the skin and other organs of the body, very red and full of blood. Why it does so we shall learn as we proceed.

Other wise men of ancient times have spoken also of wine and its bad effects. Anacharsis the Scythian said : "Wine bringeth forth three grapes, the first of pleasure, the second of drunkenness, and the third of sorrow." Demosthenes, the great orator of Greece, said that "to drink well is a property meet for a sponge, but not for a man." Seneca, a grand Roman philosopher, taught that to suppose "it possible for a man to take much wine and retain a right frame of mind is as bad as to argue that he may take poison and not die, or the juice of black poppy and not sleep."

Saint Augustine, speaking of the bad effects of wine-drinking, declares it to be "the mother of all mischief, the root of crimes, the spring of vices, the whirlwind of the brain, the overthrow of the sense, the tempest of the tongue, the ruin of the body, the wreck of chastity, a loss of time, a voluntary rage; a shameful weakness, the shame of life, the stain of honesty, and the plague and corruption of the soul." Pliny the Younger, a great writer of natural and general history, relates that King Antiochus having forced his minions at a banquet to take an excess of wine, they killed him; from which story he drew this moral: that if we tempt others into error, the consequences will fall back on ourselves.

Questions on Lesson X.

1. Name some of the wise ancient men who have spoken about wine.
2. Repeat the proverbs of Solomon on this subject.
3. What was the saying of Demosthenes about drinking?
4. What was the teaching of Seneca?
5. Repeat the saying of St. Augustine on wine-drinking.
6. What anecdote is told by Pliny relating to wine? What is the moral of the story?

LESSON XI.

DISTILLATION.

I HAVE said that those who discovered wine thought that it was a fluid quite distinct from water and from everything else. For these reasons some who were very fond of wine and of the excitement of mind it produces called it nectar, while others said it was the drink of the gods. Such vain things does this seductive drink lead vain men to declare.

For centuries on centuries wine held its special position as a drink. At last, in the middle, and as some call them, the dark ages, but which in many respects were very bright ages, an invention was introduced, or re-introduced, that made a wonderful change in science and knowledge. The invention was distillation. To distil is to make a fluid fall in drops.

The term is from the latin *distillo*, meaning to drop down by little and little. In Nature the process of distillation has been in progress always. When the earth is saturated with water, the sun, by its heat, makes the water rise from the ground in the form of vapor. The water condenses in the air as cloud, and the cloud falls as drops of water. This is natural distillation, and in time man, who in everything and every work imitates Nature either knowingly or unknowingly, at last imitated the distilling process. He put fluids of different kinds into the lower part of a vessel which he called an *alembic ;* he placed the vessel on a fire : he made the fluid rise in vapor, and then he caught the vapor, cooled it, and let it fall in fluid drops into a *receiver.* I dare say he did this first with water, but in time he applied the process to other fluids than water : and at that time many wonderful discoveries commenced.

In this line of work the chemist soon found that different fluids boil and pass into vapor at different temperatures, and that some solids can be made to give off vapors which will condense as fluids. Next he found that different vapors condense and fall down in drops at different temperatures ; and lastly he found, what would be to him the most singular fact of all, that some fluids which he thought were simple fluids, that is to say, some fluids which he thought were composed of one substance alone, are in fact made up of more than one substance, and that the different substances can be separated into distinct parts by this simple method of distilling.

When, for example, the old chemist put on the fire a fluid which had in it two distinct fluids, but so commingled that to all ordinary appearances they looked like one, he would find that the first fluid which came over in distillation was a lighter fluid than that which came over last. He would also find that the first fluid did not require so much heat to drive it into vapor, and he would find that it required more care to bring it back in the form of liquid, into his condenser. He would taste the fluid that came over first and detect it to be quite different from that which he had put into his vessel on the fire. He would taste the fluid which came over last, and which required the greatest heat, and would detect that it differed not only from what he originally put into the vessel on the fire, but also from that which came over first. Having found out these facts, he would give names to these different products of distillation. To all the lighter ones he gave the name of "spirit," because of their great lightness. Some he called ethereal spirits, they were so refined, so like the ethereal air. Others he called spirits of the substance from which he distilled them, which names are retained, as so-called vulgar terms, to the present day.

Questions on Lesson XI.

1. What fanciful ideas did the ancients hold respecting wine?

2. Describe the process in science called distillation. State the age of its discovery and why it was called distillation.

3. Give an illustration of natural distillation.

4. How did man proceed in the process of distillation; and what were the names given to the vessels he employed in the process?

5. What further discoveries came from distillation, bearing on the composition of the different substances submitted to distillation?

6. What name did the chemists give to the lighter liquids which were carried over in distillation?

LESSON XII.

SPIRIT OF WINE.

We come now to a subject relating to distillation, which is to us the most interesting of all. In the processes of distilling it was very natural that such a common fluid as wine should be experimented on in order to see what would be the result. It is said that this experiment was first made by an Arabian chemist, named Albucasis. Albucasis is believed to have lived in the eleventh century of the Christian era, and many other names have been given to him. He has been styled, shortly, "Casa," also Alsaharavius, and Benaharezerim, but he is best known in history as Albucasis.

The result of putting wine into the alembic led to the distillation of a refined light and strong spirit called, because of these properties, *spirit of wine*. It was also found that after this refined spirit was separated by the distillation another fluid remained behind, which turned out to be nothing more than water. Thus it was discovered that wine was not the simple and distinct thing it had so long been supposed to be, but was water after all, with something else added to it, namely, a spirit that can be separated by the simple processes of driving it out of the water by heat and condensing its vapor into little drops by cold.

It was no doubt considered to be a marvellous discovery to produce this spirit of wine, and very soon new discoveries were made from it. Some other cunning chemist, whose name is quite unknown to us, put the green, glassy-looking crystals which he called vitriol into his retort, distilled over at a great heat, and obtained vitriolic spirit, a very strong acid better known now as sulphuric acid. This acid was poured on common salt, and by distillation what was called the spirit of salt, which we now call muriatic or hydrochloric acid, was obtained. Thus we derive many names, still in common use, the origin of which we do not understand, until we learn why such names were first applied.

The same strong vitriolic acid was poured on spirit of wine, and the two were distilled together. By this means a much lighter and finer fluid than spirit of wine was discovered. The new fluid is so light that when it is poured into the palm of the hand it boils as briskly as water boils on a quick fire, and goes into so transparent a vapor that it seems to be lost alto-gether. It is not lost, it is only diffused or spread out in the air. But the earliest discoverers of this fine light liquid thought it was lost in the air, and owing to its exceeding light-ness and airiness, they compared it to the lightest substance they could possibly imagine, namely, to that supposed fluid which fills the spaces between the stars, and which had been called ether. They named, therefore, this fine fluid ether, and the name is still used as that by which it is best known. To this day we make ether by the original plan, viz., by acting on spirit of wine with an acid, and though we have several ethers which the earlier chemists had not, we follow their method of production.

For a long time after spirit of wine was well known, it re-mained in the laboratory of the chemist, and was there held as a kind of secret treasure, by the use of which the most astonishing experiments were carried out. It was detected that spirit of wine would dissolve very many substances which water would not touch, such as oils, resins, gums and balsams. It was detected that spirit of wine would preserve many substances from decay, such as skins of animals, and even the flesh of animals. It was observed that spirit of wine would burn, and that it gave out a flame which did not yield smoke, and might consequently be used for many purposes of heating where great cleanliness was required. These were useful applications of the spirit of wine. It would have been well if no worse use had ever been found.

Questions on Lesson XII.

1. What did the old chemists do to wine after the discovery of distillation?

2. What chemist is said to have first distilled wine? And to what age did he belong?

3. What name was given to the fluid distilled from wine?

4. What singular fact was now discovered as to the composition of wine?

5. What other new fluid was obtained from spirit of wine?

6. What were the first important uses to which spirit of wine was applied?

LESSON XIII.

ARDENT SPIRITS.

The good which came from the discovery of spirit of wine was soon tinctured with evil, by the use men began to make of the spirit as a strong drink to be taken instead of wine. They called it *vinum adustum*, burnt wine ; *spiritus ardens*, strong spirit ; as well as *spiritus vini*, spirit of wine. Then they began to bottle it for drink and called it *aqua vitæ*, water of life. So Shakespeare, in the play of " Romeo and Juliet," makes the nurse say, " Give me some *aqua vitæ*," and in the play of the

"Merry Wives of Windsor" he makes another of his characters use the same name for ardent spirits.

It would, indeed, have been correct if this spirit had first been called *aqua mortis*, water of death, instead of *aqua vitæ*, for assuredly nothing in this world has been the cause of so much crime, want, woe, disease, and death.

A very learned scholar, Mr. Stanford, has come to the conclusion that *aqua vitæ*, as it was called, that is, the spirit obtained by the distillation of wine, was used as a drink as early as the year 1260 of our present era. The Arabians, he thinks, taught the use of it to the Spaniards, and the Spaniards to the monks of Ireland. It thus came into use in Ireland, and acquired the Irish name for it by which it is still known in one form by the name of "whiskey." In the old Erse, or Irish tongue, it was called *usige-biatha*, which means *aqua vitæ*. In time this term was shortened into *usquebaugh*, and this again was shortened into *usige*, from which comes the word whiskey. Sometimes in Ireland this same strong drink is called *potheen*, or *poteen*. This word, *poitin*, means a small pot or still, the vessel from which the liquor was distilled, and poteen was, perhaps, derived from the Latin word *potio*, a drink.

We have now an idea of how the common liquor called whiskey came to be known; but there are other strong drinks which require to be explained in respect to their origin.

The Germans call the ardent spirit distilled from wine, or other fermented fluids, "*Branntwein*," burnt wine, from *brennen*, to burn, and so we get the term brandy; but probably, not until long after *usige*, or whiskey, was used to describe strong spirit.

In time men began to add other substances to ardent spirits in order to give a new flavor or taste. Amongst these added things were the berries of the juniper-tree, which berries were put into the fermented liquid while it was being distilled, and which yielded a volatile substance that gave a very peculiar flavor. The juniper-tree, called by the Latins *juniperus*, was known amongst the Italians as *ginepro*, and by the French as *genièvre*, an abbreviation from the Italian. Hence, Mr. Stanford thinks, came the name *gin*, to distinguish the common drink still sold so largely under that name, and which contains juniper. I have no doubt that Mr. Stanford is right on this point, but I ought to say some believe that a drink was first made at Geneva, in Switzerland, to which the name gin was applied, in short for the word Geneva. In some parts of England I have heard gin called Geneva, and in France, genièvre, the latter meaning the juniper-berry, not the town.

As the arts progressed, it became possible to obtain strong spirit from the fermentation of sugar. The Latin for sugar is *saccharum*. From this it is believed, by the learned authority I have so often quoted, comes the word which describes another common strong drink, namely *rum*. The word in this case is formed by striking off the first two syllables and leaving the last, *rum*.

Questions on Lesson XIII.

1. What names were originally applied to the spirit distilled from wine?

2. What strong drink was first named, and what was meant by the name?

3. What other names were applied to strong spirits?

4. What is the origin of the word brandy?

5. What are the supposed origins of the word gin?

6. From the fermentation of what substance is rum produced, and what is the supposed origin of the word?

LESSON XIV.

ALCOHOL.

Long after its discovery spirit of wine continued to be called by that name, or by one of the other names which have been described in a previous lesson. But some time in the latter part of the seventeenth century, a new name was given to it, by which it has since become universally known. The name is *Alcohol*.

Much learned discussion has taken place on the origin and precise meaning of this word. I find it was in use in the year 1698, for I have an old chemical book of that date in which the term is introduced, and in which it is applied to spirit of wine that has been very highly rectified, that is to say, distilled and re-distilled until it has attained to the extremest possible degree of strength and purity. The author of the book to which I refer was a chemist by the name of Nicholas Lemert, and he tells us some very curious facts about spirit of wine. He speaks of brandy as a fluid containing this spirit, and says the French call it *aqua vitæ.* He explains that the chemists have a sign or symbol for spirit of wine, which is like this, ℣ and, he also tells us why it is called alcohol in his time.

From Lemert's description, it appears that alcohol was a term intended to describe something exceedingly refined or subtile. He uses the word sometimes as a verb, and explains that when any substance is beaten into a very fine powder, so that it is impalpable, *i.e.*, when it cannot be felt rough to the touch, it is alcoholised. The same word, he adds, is employed to describe a very fine, pure spirit, and so the spirit of wine well rectified is called the alcohol of wine.

Other scholars have tried to trace out the origin of the word itself, and the most accepted explanation on this point is that the word is Arabic, A'l-ka-hol, meaning a very fine essence or powder used by the women of the East to tinge their hair and the margins of the eyelids. Afterwards, as described by Lemert, it was applied to all refined substances distilled by the heat of the fire.

The word is no longer used as it was in the first instance. It is now employed to express one thing, and nothing else, namely, the actual pure spirit which is produced by the fermentation of the grape, or of barley, or of sugar, or of other substances that will undergo fermentation, and yield this particular spirit. We never in these days use the word as a verb: we use it as a noun, or sometimes as an adjective. When we speak of the fluid pure and simple, we say *alcohol.* When we speak of fluids which contain a portion of alcohol, such fluids, for example, as gin, rum, brandy, whiskey, beer wine, perry, or cider—we speak of them as *alcoholic* drinks.

Before we conclude this lesson, we shall do well to recall to mind the various steps by which the fluid now known as alcohol was obtained. Let us, then, remember these five points :

1. The fluid containing alcohol that was first known was the fermented fluid obtained from fruits by fermentation and called wine.

2. The wine was distilled, and thereby a fine spirit was obtained, which was called the spirit of the wine.

3. When the spirit of wine was discovered, it was treated in different ways, by which spirits of different tastes, colors, and strengths were obtained, and called by different names, such as whiskey, brandy, rum, and gin.

4. Sugar and other substances than fruits, were made to yield spirit by fermentation.

5. At last the pure spirit, from whatever source it was got, was called *alcohol.*

Questions on Lesson XIV.

1. At what period was the term alcohol introduced to denote spirit of wine?

2. What was the original meaning of the term alcohol?

3. From what language is the term obtained?

4. To what is the word exclusively applied in the present day?

5. Enumerate the principal common drinks that contain alcohol?

6. State the five points of detail leading up to the employment of the word alcohol?

LESSON XV.

ABSOLUTE ALCOHOL.

Alcohol of wine and other spirituous drinks is, in its pure state, a clear transparent fluid, and is called *absolute* alcohol. It has a rather sharp or pungent odor, and when it is put on the lips, or tongue, or throat, it has an exceedingly biting or burning taste, so that it cannot be swallowed until it is largely diluted with water. In fact, until one part is mixed with three parts of water, it is not easy to drink it. Some persons can drink it less diluted with water; but these are they who have habituated themselves to the taking of it, and who are dangerously under its influence.

If we apply a light to alcohol it burns, giving out an indifferent light, but considerable heat. In burning it emits no appreciable quantity of smoke, and for this reason it is employed as a lamp, commonly called a spirit-lamp, for the purpose of heating without blackening. The chemists use the spirit-lamp very much in the laboratory for boiling fluids in glass vessels. The vessels remain quite clear of soot where the flame reaches them, and so the heat is not interrupted on its way to the liquid by a layer of soot, which, if present, would seriously cut off the heat. It is easy, by experiment, to prove this by trying to boil water with a spirit-lamp and with an oil-lamp. The oil-lamp may actually give out more heat, but owing to the soot that is deposited on the vessel containing the water, the operation of boiling will be much more tedious than in the case of the water subjected to the heat of the spirit-lamp.

While alcohol is being burned, some fluid is produced, which fluid is water. During the burning there is given off vapor of water. It is difficult, therefore, to dry any substance completely over the spirit-lamp, because the water that is diffused will enter the substance if care be not taken to prevent it.

The water which is present results from the combustion of the alcohol. It is made in the process of burning. How this occurs we shall see further on.

If, while alcohol is burning, we hold over the flame the mouth of a bell-jar, so as to catch whatever is rising from the jar, without interfering with the admission of air to the flame, we can condense, on the inside of the jar, some of the water that rises as vapor. At the same time we catch something else which we cannot see, but which is present in the jar in the form of a gas. The said gas differs very singularly from the common air with which the jar was filled before we inverted it over the spirit-flame ; for, if now we put a lighted taper into the bell-jar, the light quickly goes out ; or, if we draw the gas from the jar into our lungs, we feel a sense of suffocation, and should indeed be suffocated, as the taper was, if the quantity of gas in the jar were enough, and we were made to breath it for a sufficient time. Further, if into the jar, while it contains the gas which came off from the burning spirit, we pour a solution of lime, lime-water, and shake up the solution in the jar, we see trickling down the sides of the jar a milk-like fluid, and if we collect this fluid from the jar and gently dry it, we get a powder which we recognise as resembling common chalk powder in its appearance.

The gas which is given off from the burning spirit, which puts out flame, which puts out life if it be breathed, and which unites with lime and makes the chalky powder, is called carbonic acid, carbonic dioxide. It is an acid gas. It goes off from burning coal in ordinary fires, and it also goes off in our breath as a product of that animal combustion by which the heat of the living body is maintained.

Questions on Lesson XV.

1. What is the appearance of the spirit called absolute alcohol?

2. What is the odor and taste of alcohol in its pure form?

3. What effect does it produce when it is put on the lips, tongue, or throat?

4. In what peculiar manner does it burn when a light is applied to it?

5. What vapor goes off from it in the process of burning?

6. What gas goes off from it in the process of burning, and under what other conditions is the same gas produced and given off?

LESSON XVI.

ALCOHOL IN WINES, SPIRITS AND BEERS.

THE amounts of alcohol, in the form of absolute alcohol, present in different wines, as measured by the volume, varies from eight or nine to twenty-five per cent., so that in two wines on the table a glass of one may be equal, in alcoholic strength, to three glasses of the other. This fact illustrates very well how foolish it is for people to speak of wine and to partake of wine, in the way they so often do, as if wine were, like water, a fluid of uniform composition.

The strongest wines are the ports and sherries. The so-called foreign port wines are usually "fortified," as it is said, up to twenty-five per cent. of alcohol. It is a poor and unsatisfactory port which does not contain over twenty per cent. of spirit. Sherry contains often from twenty to twenty-five per cent., but as a rule it is less brandied than port, the average quantity of alcohol in it being from twenty-two to twenty-three per cent.

The wines known as British ports and sherries are weaker than ports and sherries. The wines sold as French wines are much less rich in alcohol. Their average strength may be taken at ten per cent. Champagne holds an average of from ten to eleven per cent. Moselle wine is of about the same strength as champagne.

The Rhine wines, called commonly Hocks, vary exceedingly in strength. Those who are practised wine-drinkers are given to look on hock as a kind of "water bewitched," as a wine so poor in alcohol and so rich in water as to be scarcely worth the drinking, except for some delicacy of flavor or aroma which their senses have learned to distinguish, or which they profess to have learned to distinguish, for there is about the tasting of wine a great deal of nonsense and pretension. The red Rhine wines are all of fuller strength, yielding sixteen to seventeen per cent., by volume, of alcohol; and the English home-made wines from eight to eleven. One of the home-made wines—gooseberry—so closely resembles champagne that it sometimes passes for champagne amongst those who are not versed in wine.

The amount of alcohol present in the liquids sold under the name of spirits is much larger than that which exists in wines, for which reason those who are called spirit-drinkers are more quickly injured or killed by their bad habits than are pure wine-drinkers.

The spirits in most common use are gin, whiskey, rum, and brandy, and their strength in alcohol lies in the order in which I have placed them, from the weakest to the strongest.

Gin is the weakest spirit, but it contains, nevertheless, from thirty-eight to thirty-nine per cent., by volume, of absolute alcohol.

Whiskey is stronger in alcohol than gin. It contains from forty-five to forty-six per cent. of alcohol.

Rum is richer still in alcohol. The "best" rum has in it as much as forty-eight and a half per cent., by volume, of absolute alcohol.

Brandy is the liquid which contains the largest quantity of alcohol. This liquid, in the state called "good," contains fifty-three per cent., or even fifty-four per cent. of absolute alcohol, and no brandy, probably, is tolerated that contains less than fifty per cent., by volume. Brandy is, consequently, of all the spirit drinks, the most dangerous and the most fatal.

Ales, and beers, and stouts, and porters are much less richly charged with alcohol. They approach, in this respect, the light wines, such as champagne. Some specimens of ales and stouts contain as much as ten per cent. of alcohol, and in very strong old ale that quantity may be exceeded. There is, however, a great deal of trickery played with the ale which is commonly sold in retail, so that it is difficult to arrive at any correct standard. It is a double pity that people should be so ignorant as to pay a high price for such diluted drink, when simple water would quench the thirst much better, and when a little oatmeal added to water would give much more strength wherewith to labor.

Some of the hardest labors which working men have had to accomplish have been best performed by men who have had no other drink than water, to which oatmeal has been added.

Questions on Lesson XVI.

1. What is the range of variation in the quantities of alcohol contained in the fluids called wines?

2. How much alcohol is present in the wines called port, sherry, and Madeira?

3. How much alcohol is present in what are called French and Rhine wines?

4. What is the average alcoholic strength of English home-made wines?

5. What are the spirits in most common use as dangerous strong drinks; and which of them is richest in alcohol?

6. What is the percentage, by volume, of alcohol in gin? in rum? in whiskey? in brandy?

7. What are the varying percentages of alcohol in the fluids sold as ales and porters?

LESSON XVII.

ALCOHOL AND ANIMAL LIFE.

WE are now prepared, by our previous studies, to consider in what way alcohol acts on the bodies of men and animals. In this part of our work we shall have to think almost entirely of the action of that one alcohol which is called ethylic, the second in the alcohol series, the alcohol that is found in wine and other common intoxicating drinks. We must confine our attention to this alcohol altogether ; but it sometimes happens that the other alcohols are mixed with it in drinks, either by accident or intention, and it is well, therefore, to remember what has been briefly stated in preceding lessons about them and their action. For the present purpose we have to think only of the alcohol which has been so long known, and which exists in wine, spirits, beer, cider, perry, gin, rum, whiskey, brandy, and other common liquids sold for drink, and called now, universally, strong or alcoholic liquids.*

Whenever the word alcohol is used in future lessons, it must be understood to refer to this best known representative of the alcohol series, namely, to ethylic alcohol. Whenever any other alcohol is referred to it will be defined by its name in full.

There is no animal that may not be affected by alcohol. At all events, I know of none. Some animals will swallow without injury substances that would be poisonous to man. A pigeon will take, without showing the slightest symptom, as much opium as would kill several men. A goat will swallow, without injury, a quantity of tobacco which would kill several men. A rabbit will swallow, without injury, a dose of belladonna that would kill several men. But neither the pigeon, nor the goat, nor the rabbit can swallow alcohol without being influenced by it in much the same manner as a man would be.

---

* For a description of the various kinds of alcohol, and their composition and character, see the appendix.

41

To show how universal is the action of alcohol on living animals, a fact bearing on the matter of weight of the animal is important to remember. I found that the dose of alcohol which is sufficient to affect an animal bears a relation to the weight of the animal, so that a knowledge of the weight of an animal will tell how much alcohol is required to produce a certain effect. This only holds good, however, in respect to animals that are not habituated to the use of alcohol. As soon as the taking of alcohol becomes a habit, either in a lower animal or in a man, the quantity required to produce obvious effects has, as a rule, to be increased.

And here I may stop for a moment to point out a very important practical lesson, derivable from what I have just stated. The lesson is, that alcohol taken into the body causes sensations and effects which soon seem to become a necessary part of life, but which can rarely be sustained from day to day in any person without leading to an increase of the dose which first caused the sensations or effects. A man takes other foods and drinks without experiencing this change. If he takes a pint of milk, or a pint of water, or half a pound of meat to-day, he does not find a desire to go on day after day increasing those quantities of milk, water, or meat indefinitely, until one of them takes the place of almost everything else, becomes the only thing really cared for, and at last produces symptoms of the most terrible disease. The taking of alcohol does, however, lead to this kind of habit in respect to itself ; it excites a constant craving for more of itself, more and yet more, day after day, until at last the deluded man or woman that is used to it may increase the quantity until, caring for nothing else whatever, he or she becomes the victim of the most terrible and fatal of bodily and mental diseases.

Questions on Lesson XVII.

1. What is the peculiarity of the action of alcohol on the different classes of living animals ?

2. Is this action peculiar to all poisonous substances ?

3. Give examples of the inaction of some poisons on particular animals.

4. What fact respecting weight shows the similarity of the action of alcohol on different kinds of animals ?

5. How does alcohol differ from simple foods in regard to its satisfying effect on the living body ?

6. What is a common effect of alcohol on those who take it habitually, day by day, in regard to the quantities taken ?

LESSON XVIII.

ALCOHOL AS A FOOD.

It matters very little how alcohol is put into the body to secure its absorption into the blood; wherever the blood can carry it there it goes, and those parts of the body which hold the largest quantity of water keep it longest in combination. If it were injected through a fine hollow needle under the skin, it would find its way all over the body, and it very readily finds its way from the stomach after it is swallowed. From the stomach it is absorbed quickly into the veins, and, mixing with the blood in them, is carried by the circulation into every vascular organ and part.

Before there can be any absorption into the body, however, one thing must be done. The alcohol must be well diluted with water. If the alcohol be brought into contact with the living structures of the body in the pure, or, as it is usual to say, in the neat form, it has such an affinity for the water in those structures that it seizes the water and renders the parts hard and dense. There are two terms used in scientific language to indicate the change that is thus brought about. One of these terms is "coagulation"; we say that alcohol "coagulates" the tissues or fluids. The other term is "pectous"; we say the alcohol renders the tissues or parts "pectous," which means viscid or thick. Alcohol acts on the blood, on all the membranes of the body, and on all the fluids which contain albuminous matter, in the manner now described when it is not freely diluted with water. Brandy, which consists of about equal parts of alcohol and water, is sufficiently strong to produce this coagulation.

It is not very probable that Nature, so wise and good, would give to man a substance for a common food which is so easily destructive in its action unless it be taken with the utmost care. She did not give man boiling water for drink, and leave him to discover that the water was dangerous, unless it were cooled to a certain safe degree ; neither did she give him such active chemical fluids as ardent spirits for his nourishment. Yet there are many persons who declare that alcohol is a food, and that it is given to us for our use as a food.

Such persons do not think, when they make the statement, what a food is. We know from the best of teachers, from Nature herself, what a true food is. We know that milk is a true standard food. Milk gives us water for the purposes it fulfils in the body ; it gives us caseine to supply new structures, muscles and other active organs; it gives us salts for building up the skeleton and other uses; it gives us fat (butter,) and sugar to produce animal warmth and power. (See Lesson VII.)

Everything is clear enough about milk as a food : we can drink milk without being burned, and we can see where all its parts go to, and what they do in our bodies. But when we come to look at alcohol, we can see no such qualities or uses for the purpose of food of any kind. It does not supply water : therefore it is not a water food. It does not supply salts : therefore it is not a saline food, and would never help to make the bony skeleton. It does not supply caseine, albumen, fibrine, or any other of those substances which go to build up the muscles, nerves and other active organs. The foods which supply these parts contain nitrogen and phosphorus and sulphur as necessary elements, and we are sure that alcohol does not contain those elements ; it is not, therefore, an active structure-building food, and unless it be taken in combination with sugar, it does not even make fat.

A substance that neither gives water nor any substance for building up the body cannot be a nourishing food. Some, however, will urge that it gives warmth and strength, and is in that sense a food. We shall see, step by step, that this too is a delusion ; we shall see that alcohol makes the body cold and makes the muscles weak.

Milk is a drink, and milk is a standard food sent by Nature. Compared with milk, alcohol shows no trace of being a food in any particular. We want no better evidence, and we could have no better proof, that alcohol is not a food.

Questions on Lesson XVIII

1. How will alcohol enter the structures of the living body, and what is the medium or fluid through which it is diffused ?

2. What must be done to ensure the diffusion of alcohol through the living body, and what happens if undiluted alcohol is brought into contact with the tissues of the body ?

3. Why would a fluid possessing the properties of alcohol be dangerous as a food ?

4. What are the qualities of a natural standard food, such as milk ?

5. In what respects does alcohol differ from such a standard ?

6. Sum up the reasons why alcohol is not to be considered a food.

LESSON XIX.

ALCOHOL IN THE BLOOD.

WHEN alcohol has been introduced into the body, it is very soon absorbed into the blood, and by means of the blood is carried all over the body. It enters the blood by those vessels which are called veins, and which convey the whole stream of blood from every part of the system into the right side of the heart. The veins are the bearers of blood, which, having traversed all the organs, and supplied food to all parts, and been the means of supplying warmth, is now returning to the heart, picking up nourishment by the way and hastening to the lungs, into which it is lifted by the stroke of the right side of the heart, in order to give up in the lungs a charge of carbonic acid, with which it is loaded. The veins are the carriers of blood to the heart, and when alcohol is present they carry it mingled with the blood that is in them.

As the blood from the veins passes over the lungs, it gives up in its course its carbonic acid gas, and in place of that gas, or I should say in exchange for that gas, it receives oxygen gas from the air which we take in at each inspiration. The plan by which the blood is enabled to give out carbonic acid and to take in oxygen is very simple and beautiful, and it is necessary for my purpose that in a few words the plan should be de-

scribed.   In the blood there float millions of little round bodies
called   red blood-globules, or blood-corpuscles, or blood-discs.
These are so small they can only be seen by means of the
microscope :   indeed, they are so small that if we were to
measure one of them across, as we would measure a sixpence,
we should find the diameter to be no more than the three
thousand five hundredth part of an inch.   This smallness is
necessary for two grand purposes at least.   It is necessary in
order to allow the corpuscles to pass easily through the minute
blood-vessels of the body, and it is also necessary in order to
expose a large surface of these corpuscles to absorb the gases
which it is their duty to pick up and carry.   When the blood
in the veins is floating towards the right side of the heart,
which communicates with the lungs, it carries with it the carbonic acid, and, as I have found by experiment, a great part
of this gas is condensed in these little bodies, the corpuscles.
Arrived at the lungs, the blood comes into such contact with
the air we breathe at each breath we draw in, that the oxygen
gas in the air is freely absorbed by the little corpuscles, while
the carbonic acid is given up into the air-passages of the lungs,
and is thrown off by the breath with every breath we throw
out.   In this process the blood changes in color.   It came
into the lungs of a dark color ; it goes out of them a bright
red.   It makes its way, after the change that has occurred,
into the left side of the heart, charged with oxygen gas.   The
left side of the heart contracting upon it, it is carried into a
set of vessels called arteries, and through them, with a steady
stroke, due to the contraction of the heart, into all parts of the
body.   The blood is now called arterial, or red, or oxygenated
blood.

Questions on Lesson XIX.

1. How does alcohol find its way over the system ?

2. What is meant by venous blood ?

3. What is the use of the red blood-corpuscles ?

4. In the lungs what do these minute bodies give up, and what do
they take up ?

5. Where does the blood go after it leaves the lungs ?

6. What is the blood called after it has left the lungs, and in what
particulars does it differ from venous blood ?

LESSON XX.

ACTION OF ALCOHOL ON THE BLOOD.

In addition to the blood-corpuscles of the blood to which the name of red corpuscles is applied, there are, in much smaller numbers, other corpuscles called white. These white corpuscles are larger than the red, and it is now believed that the red corpuscles are derived from them. There is also in the blood, distributed through it, a jelly-like substance called fibrine. The fibrine is present in the proportion of about two and a half parts in the thousand, and when blood is drawn from the body the fibrine becomes partly solid. It goes into a clot, or as it is said by scientific men, it coagulates. The process of coagulation of blood is of great importance in life. When a person cuts his finger so as to make it bleed, clotting or coagulation of blood takes place, and by that means the flow of blood is stopped. If such clotting were not to take place, the bleeding would continue until the whole of the blood drained out of the body even by a small opening. In some instances people have died from this cause, their blood being poor in fibrine. The fibrine has still more important uses. To use a common phrase, it holds the blood together, so that when the blood reaches the extreme vessels it does not easily transude or pass through the porous structures of the body, and, except under proper regulation for the purpose of building up the body, it does not leave the vessels at all. If the blood be deficient in fibrine, or if the fibrine from any cause be held in too fluid a state, then the blood does exude through the vessels and makes blotches on the skin and other membranous parts, and indeed, in such instances, it sometimes flows from the body in some parts, as from the gums. We say of persons in this state that they are scorbutic, or that they have scurvy, a painful and enfeebling disease.

Lastly, albumen, salts, and water, are parts of the blood ; the water making up seventy-nine parts in the hundred.

I have stated these few facts about blood in order to explain more perfectly the action of alcohol upon it. When

47

alcohol is made to enter the body by any channel it finds its way into the blood. In the blood it spreads or diffuses through the water, and is carried, mixed with the water, over the body at large.

In the smallest quantities diffused in blood, alcohol is a foreign substance, doing there no good certainly, and fulfilling no purpose that could in any sense be called natural. When the quantity of it is small, so small as one part in five hundred, it is injurious; while over and above that quantity it is a source of serious derangement, and it may be of danger.

The parts of the blood on which alcohol acts injuriously are the corpuscles and the fibrine. The red corpuscles are most distinctly affected. They undergo a peculiar process of shrinking from extraction of water from them.

They lose also some of their power to absorb oxygen from the air. In confirmed spirit-drinkers the face and hands are often seen of dark mottled color, and in very bad specimens of the kind the face is sometimes seen to be quite dark, almost of the color of the skin of a mulatto.

In these drinkers the process of oxidation of blood is impeded to an extreme degree by the presence of the alcohol. They always are very sensitive to cold, and in the winter weather they are subject more than most other people to affections of the lungs which are serious in character. They die, in fact, in great numbers, under such conditions of the air, for their vital organs, their lungs, liver, kidneys, brain, are like their skin, congested readily, first because the blood-vessels are deficient in power, and the course of the blood through the vessels is readily checked; secondly, because the blood itself cannot take up the vital air in the natural degree.

Questions on Lesson XX.

1. Name the other parts of the blood beside the red corpuscles.
2. What is the average amount of fibrine in the blood?
3. What phenomenon occurs when blood is drawn from the body?
4. What are the uses of the fibrine in the blood?
5. What parts of the blood are specially affected by alcohol?
6. What are the effects produced by alcohol on the red corpuscles of the blood, and how are those effects shown on the body of the hard drinker?

LESSON XXI.

FURTHER ACTION OF ALCOHOL ON THE BLOOD.

THERE is another action of alcohol on the little red corpuscles of the blood. When it enters the blood in large quantities, it causes the corpuscles, rendered smaller and irregular from loss of water, to adhere together in masses. This change is a cause of serious embarrassment and danger. When it occurs, the masses of corpuscles prove directly obstructive to the course of the blood in the minute or hair-like vessels of the body, through which vessels the corpuscles naturally pass in file. The obstruction which is produced by the massing together of the corpuscles is directly mischievous. It leads to congestion of blood in the organs in which it takes place, and sometimes it produces disease at the part where it occurs. It has also the effect of preventing a due absorption of oxygen by the blood in the lungs.

The bad action of alcohol on the blood is not confined to the red corpuscles; it extends to the fibrinous or plastic part of the blood. If the alcohol be taken in very great excess indeed, it causes in the blood coagulation of the plastic part or fibrine. The fibrine, under these circumstances, becomes solidified in the blood-vessels, and, so charged, stops the current of the blood in its course through the vessels. There have been instances in which death has been the almost instant consequence from a change of the blood as here described. Some foolish man for a wager, or other silly motive, has swallowed, straight off, a large draught of strong spirit, and thereupon has died on the spot. In instances of this nature the blood may be affected by the poison in a few seconds, and the whole of the blood in the heart be made a semi-solid mass or clot, by which the circulation is absolutely stopped, and death is rendered inevitable.

I do not assume that many people, however fond they may be of alcohol, choose to kill themselves with it in this direct manner. If they did, their act would be styled suicide, and the attempt to perform the act would be contrary to the law.

49

But there is another somewhat similar accident, which a fair number of persons, in their madness for alcohol, do carry out. They so charge their blood with alcohol, that in the minute blood-vessels minute masses of fibrine are coagulated into clots, which cut off the current of blood from little parts of the organs of the body, whereby those organs are injured. Sometimes these tiny obstructions occur in the brain, and so the beginnings of paralysis are produced. In other instances the obstruction to the blood in the vessels, owing to the coagulation of the fibrine, occurs in the liver or the kidney, and so those organs are injured, and, in time, are made seats of fatal diseases.

The signs of injury in the person in whom this accident occurs are of various kinds, varying according to the place in which the accident has happened. The signs may be those of giddiness when some small centre of the brain is affected ; or they may be those of palsy, owing to the nervous communication with the muscles being for the time cut off ; or they may be those of injury to one or other of the senses, owing to the nervous mechanism of one of the organs of the senses—the eye or the ear—being injured. In extreme examples of this accident, the brain is so much injured, and the circulation through the brain so seriously stopped, that the body falls into an unconscious state, as if it had been knocked down. It is then said that *apoplexy*, which means literally striking down, has taken place. Great numbers of hard drinkers die from the cause here named.

When the accident of obstruction of blood occurs, in a similar manner, in the vessels of the lungs, liver, or kidneys, the signs of disease are presented in those vital organs, and life is endangered from the failure of their natural functions.

Questions on Lesson XXI.

1 What further injurious action does alcohol exert over the red corpuscles of the blood ?

2. What injury to the body results from this change ?

3. By what effect on the blood may alcohol cause sudden death ?

4. By what effect on the fibrine of the blood may alcohol injure parts of vital organs ?

5. What signs of injury to the body may follow the changes thus caused ?

6. What is the name given to a common disease from which hard drinkers suddenly die, and what does the word mean ?

LESSON XXII.

DISEASED BLOOD FROM ALCOHOL.

IN some instances, as we have seen, alcohol causes thickening of the blood in parts of the body. But in most persons who indulge freely in alcohol, the blood is thin, and easily flows from the blood-vessels when they are wounded by a cut or tear. The alcohol, from its great affinity for water, induces those who drink it to imbibe water or watery fluids to an excess. Those men of great size of body, who are called draymen, and who too often drink ale and porter in very large quantities, are singularly subject to danger whenever they suffer from accident. On such men, when they are greatly addicted to drinking of malt liquors or other spirituous fluids, the surgeon looks with the greatest anxiety, if he has to perform on them even trifling operations, for he knows they are liable to dangers which are frequently of a fatal character.

Alcohol, when it is freely diluted, acts, as I have already said, on the fibrine of the blood, rendering the fibrine unduly fluid. In fact alcohol acts on the blood in somewhat the same manner as salt does, and gives a tendency to a disease similar to that which is produced by living a long time on salted provisions, to which disease the name of scurvy is commonly applied.

In the different attempts that have been made to get near to the North Pole, the disease known as scurvy has often been developed in the crews. Salted foods and exposure to cold have, as a rule, been described as the causes of this disease, and, without doubt, they play an important part. But the truth is now clearly out that the ration of grog, as it is called, in other words the ration of some spirit, such as rum, which is allowed the sailors, adds very greatly to the evil. In the latest Arctic expedition, the sailor who went through the severest trials, who escaped scurvy, and who held his own all through in the healthiest form, was a brave man named Adam Ayles ; and Adam Ayles, from the beginning to the end of the voyage, never touched a stimulant to drink of it. On the other hand, those who took a moderate excess of alcohol suffered severely, and one man, whose likings for the poison were too strong for his reason, had to be kept from every exposure to cold, and in a mere half-useful position during a great part of the voyage.

In scurvy, the blood is so fluid that it may pour out of the small blood-vessels at some points of the body without those vessels being wounded mechanically. It may transude into the structure of the skin, and produce deep blotches of the skin like bruises, and even blotches on the internal membranes and organs of the body.

A disease of a similar kind sometimes affects poor people who live on land, and who are obliged, for the sake of economy, to subsist on coarse bread, bacon, or other salted foods, and who try to make up the deficiency by resorting to beer, or perhaps to stronger drinks, such as gin. These people become what is called " scorbutic," and much weakened. They feel all the changes of weather severely, and are great sufferers during the cold of winter. If they would spend the money which they throw away on the drinks that are so hurtful, and would take to plain, nourishing foods, like oatmeal and pease-meal, they would be very much better off in mind, body, and estate. Two tablespoonfuls of oatmeal with one of peasemeal, made into a liquid with milk and boiling water, and flavored according to taste with salt or with sugar, forms a drink worth any number of glasses of ale or other alcoholic fluid. Such a drink does indeed maintain the strength and keep out the cold.

Questions on Lesson XXII.

1. What effect does alcohol have in modifying the amount of water in the blood ?

2. What is the name of the disease that is produced by living too long on salted foods ?

3. What does alcohol do in regard to this disease?

4. When alcoholic drinks are taken in excess with water, what dangers are incurred?

5. Give an illustration of a class of men who run great risks from this cause.

6. Give an example of a simple and cheap drink that is better for sustaining strength and keeping out cold, than beer or other spirituous drink.

LESSON XXIII.

RESPIRATION AND ALCOHOL.

WE have learned that there are two sets of blood-cells floating in the blood—the red and the white cells. We have also learned that the red blood-cells are the most numerous. We have learned that the duty of the red blood-cells is to receive and condense the oxygen of the air during the process of breathing or respiration, and to carry that oxygen into all parts of the body by means of the circulation. We have learned that the oxygen so carried is given up in the tissues for the support of the combustion of the body and the supply of the animal heat. Lastly, we have learned that in this process of combustion a new gas is produced which is called carbonic acid gas, and which is brought by the venous blood, to escape in breathing, and be replaced by oxygen.

We are fully aware that if all the steps of this simple process are not duly carried out, the body is not in health. If anything whatever interferes with the proper reception of oxygen by the blood, the body is not properly oxidized, the animal warmth is not sufficiently maintained, and life is reduced in activity.

If for a brief interval of time the process of breathing is stopped in a living person, we see quickly developed the signs of difficulty, and we say the person is being suffocated. We observe that the face becomes dark, the lips blue, the surface cold. Should the process of arrest or stoppage of the breathing be long continued, the person will become unconscious, will stagger, and fall, and, should relief not be at hand, he will in a very few minutes die.

Some substances which we take into the body, and which
find their way into the blood, have the effect of checking the
process of absorption of oxygen by the blood corpuscles, and
also of stopping that vital process of oxidation in the extreme
parts of the body by which oxygen is made to combine with
carbon, with the production of animal heat.

I found by experiment that in presence of alcohol in blood
the process of absorption of oxygen was directly checked, and
that even so minute a quantity as one part of alcohol in five
hundred of blood proved an obstacle to the perfect reception
of oxygen by the blood.

It is not difficult to explain why alcohol should interfere
with the function of the red corpuscles when it is present in
very large quantities, because in large quantities it produces
extreme changes of a physical character in the corpuscles,
which must be injurious. The corpuscles are reduced in size,
and they are made so irregular in shape that it is almost impossible to recognize them so as to say to what class of animal
they belong. In such a state it is not to be expected that they
can discharge the duties of their office well. But when the
quantity of alcohol in blood is so small as not to affect the
blood corpuscles visibly, in the manner described above, it is
still capable of interrupting the entrance of oxygen, although
not to so marked a degree.

I must not enter into any speculation as to the cause of
this interruption : it is sufficient to state the simple fact that
alcohol does interfere with the process by which the blood is
properly charged with the vital air that passes to it by the
lungs in the act of breathing.

The evil from the presence of alcohol in the blood does not
end merely with the absorption of alcohol. The presence of it
modifies the combination of oxygen with many substances that
otherwise readily combine with it. Added to dead animal
substances alcohol prevents that combination of oxygen with
the elements of the dead matter which ends in putrefaction or
decomposition, and hence alcohol is said to be an antiseptic—
that is to say, a substance which prevents putrefaction or
decay. It plays the same part in living organs of the body
when it is carried to them by the blood. Thus alcohol interferes with that part of the animal function which provides for
the application of the air we breathe to the wants of the body.

Questions on Lesson XXIII.

1. Trace the steps which lead to the production of natural warmth in a living animal.

2. What happens if for a short time the process of breathing be in any way checked?

3. What effects do some substances, when they are taken into the body, have on the processes leading to production of animal warmth?

4. What effect has alcohol in this respect?

5. What effect has alcohol on the blood in respect to the process of breathing?

6. What effect has alcohol on the blood in the extreme parts of the circulation?

LESSON XXIV.

STANDARD ANIMAL WARMTH.

WE may gather from the last lesson that alcohol in any drink—in wine, in beer or ale, in spirit—is not a substance that can be used in order to keep the body of a man warm. At the same time many persons, perhaps most persons, who have lived in countries where strong drinks are taken, have really believed that one of the reasons why alcoholic drinks should be used for drinking purposes is, that alcohol warms the body. This idea finds expression in a great number of ways. People commonly say to their visitors in winter, "Take another glass to keep out the cold;" and although they show their own folly in summer by saying, "Take another glass to keep yourself cool," the time which passes between the two sayings is quite sufficient to hide the fact that one or other of the sayings must be silly, and can only be explained away by an argument that strong drinks warm the cold body and cool the heated.

In simple truth, however, the sayings I have quoted, and many others of a similar kind which I could quote, are mere sayings which pass from mouth to mouth without any real sense. They are spoken from habit, not from thought, and are very much like the sayings of a parrot, which sound sensible enough, but do not in any way prove that the bird tells anything that it means or understands. About alcohol, and its influence in keeping out the cold, no one stayed to inquire until quite lately. Some few very wise men have long suspected that the idea was an error, and the experiences of those who have been exposed to extreme cold have led them to the same conclusion ; but the precise fact has not, until late years, been proved and made perfectly clear.

We now have the most positive proofs that alcohol is an agent which impairs and even arrests the process by which the body is made and kept warm by natural supplies of heat. Quite apart from the reasons why it should impair or arrest this process, which reasons have already been hinted at, we are now aware that it does so, and the evidence that it does so cannot be too well or too widely known. It fell to my lot to make this truth evident, and to place it on record, from the best of all evidence, that of experiment.

I will tell in a few sentences what the evidence consisted of, and what it defined.

In all animals there is a certain fixed or natural temperature, as can be proved over and over again by experimental trial. If we take a fine thermometer, that is, the instrument with which we measure heat, and place it under the tongue of any man, woman, or child, the thermometer will tell us that the natural temperature in every case is a little over ninety-eight degrees by Fahrenheit's thermometer, which is the one most in use in this country. If in any case we find the temperature to be much above or below this degree of ninety-eight, we are obliged to come to the conclusion that the person on whom the observation is made is, for the moment, out of health ; and very often by this one sign alone of temperature or warmth physicians detect the most serious forms of disease. So we say that for man ninety-eight degrees is the natural standard temperature. Whether the man be in a hot country, on the equator, or in a cold country, near the North or South Pole, the standard is, with very small variations, the same

This fact about a standard animal temperature is a fact to be kept very clearly in mind. The perfect life can only be maintained under equal temperature of body. A variation of ten degrees on either side of the standard is fatal if it be at all prolonged. In hot climates men require less, in cold climates more, food for production of heat.   In hot climates men cool by evaporation of water from the body, that is, by perspiration. In cold climates, to retain the animal heat, they clothe themselves thickly.   In hot climates they absorb least oxygen, in cold climates most, from the air.   By these means the animal warmth is equalized.

Questions on Lesson XXIV.

1. What is the common belief as to the influence of alcohol on the warmth of the body?

2. What are common forms of expression in winter time, bearing on this topic?

3. What has led to such expressions?

4. What is the fact as to the truth or falseness of such expressions?

5. What variations from the natural temperature of the body are fatal if prolonged?

6. How is the standard temperature maintained in different parts of the world?

LESSON XXV.

ANIMAL LIFE UNDER ALCOHOL.—THE FIRST STAGE.

THE standard of temperature is not the same in all animals, though it is the same in every animal of the same species.   In man and the higher animals, the animal temperature is so high that they are called warm-blooded.   In other animals, frogs, for example, the temperature is, by comparison, so low that they are called cold-blooded animals.   But still in the different classes the temperature is the same in each class during health.

The temperature of man is a little above 98° Fahrenheit, in the horse it is 100°, in the ox 101°, in the dog 102°, in the cat 102°, in the guinea-pig 102°, in the rabbit 104°, in the sheep 104°, in the goat 104°, in the fowl 108°, in the pigeon 109°, in the duck 108°.   These are all warm-blooded animals. In some cold-blooded animals the temperature may be as low as 57° Fahrenheit.   It is 57° in the leech.   In the turtle it is 86°, in the frog 70°.

For the perfect life of every one of these animals, and as far as we know, of all animals, steadiness of bodily temperature is essential. By accurate measurement of the temperature it is easy to tell what agents or things acting upon the body will cause the temperature or warmth to vary from the natural standard. To ascertain what part alcohol plays on the animal warmth was a task I had to perform.

When I began to follow out this task I did not know for a certainty what part alcohol plays, but I was rather impressed with the common idea that it makes the body warm, for I knew that after taking wine the faces of people become red and flushed, and that such persons look warm and feel warm whether they are so or not.

The subject was of such great importance that it took me more than three years to pursue it, and it would take longer space than this book altogether contains to relate the many trials that were made. I must therefore be content to relate the results or conclusions in a very few sentences.

To begin with the simplest fact. I found that when a small quantity of alcohol is taken, just a sufficient quantity to produce a decided effect on the body, an effect that may be felt by the person who takes it, and which gives evidence of action to another person who is looking on, there is caused a certain short stage of action, to which I give the name of the first stage of alcoholic influence.

In this stage in man, the person who has taken alcohol feels exhilarated or freshened. He says he is brightened or cheered, that he feels in a kind of glow, that his thoughts seem to flow more quickly, and that he is warmed from head to foot. If you look at the man you see that his face is red and flushed, that his eyes look bright, that in manner he is a little excited, and that he talks more quickly or freely. If at this stage we take the temperature of the surface of the body, or of the mouth, we find that the temperature or warmth is actually raised. In those who are not much accustomed to alcohol, the warmth may be raised half a degree; in those who are accustomed to alcohol, the warmth may be raised a full degree, or even a degree and a half Fahrenheit, beyond the natural standard. The statement, therefore, made by the person, of feeling so much warmer, is quite correct. The skin of the body is the most sensitive of surfaces, and the sense of warmth through or over the whole surface of the skin is conveyed from it to the brain and nervous centres of the body, by which we are enabled to feel. It seems, therefore, as if warmth were communicated. When we stand before a fire and get a sense of warmth by that means, the impression of the warmth is made on the skin, and is conveyed in a similar manner to the hand. When we burn the skin, the warmth is the sharpest of all pains.

Questions on Lesson XXV.

1. What is the range of the standard of warmth in animals of different classes?

2. What do you mean by a warm-blooded, as distinct from a cold-blooded, animal? Give an example of each.

3. State the degree of warmth of the bodies of the following animals: Man, the horse, the dog, the rabbit, the pigeon, the frog.

4. What is the effect of alcohol on the body of man in the first stage of its action?

5. What is the appearance of the skin during this first stage?

6. What is the state of the body in respect to its surface warmth during this stage?

LESSON XXVI.

THE FEELING OF WARMTH FROM ALCOHOL.

THE first stage of action of alcohol on the body is the same in other classes of animals as in man. At all events, this is the fact so far as research has progressed up to the present time. In all animals that have been observed while they are under its influence, alcohol, in its first action, causes the external warmth of the body to rise. In birds the warmth increases a degree, and in other animals at least half a degree Fahrenheit.

Well, then, some one will say, the old notion that alcohol warms the body is quite correct. We do take alcohol to keep out the cold after all! Not quite so fast. When I burn my

finger it feels intensely hot, or if I hold my hand in front of
a fire until I can scarcely bear it, then it feels hot to me, any
one else touching it would say it was hot, and a thermometer
applied to it would show it was hotter than is natural. But
for all that my body is made no warmer. If I go out in the
cold and pick up snow, or plunge my hands in ice-cold
water, when I take them out they soon begin to look
very red, they soon begin to burn, and I say I have the " hot-
ache," and if I didn't know better, I should declare I was
heated. But my body would not be made warmer altogether
in either case.

The facts are just the same with respect to the sense of
warmth caused by alcohol. It is a *sense* of warmth that is
felt, not an actual warmth that is given to the body. The
sense of warmth occurs in the following way : When the alcohol enters the body, and by the blood-vessels is conveyed to
all parts of the body, it reduces the nervous power of the small
blood-vessels which are spread out through the whole of the
surface of the skin, and all the surfaces of the body. In their
weakened state these vessels are unable duly to resist the course
of blood which is coming into them from the heart under its
stroke. The result is that an excess of warm blood, fresh from
the heart, is thrown into those fine vessels, which causes the
skin to become flushed and red, as it is seen to be after wine or
other strong drink has been swallowed and sent through the
body. So, as there is now more warm blood in the skin than
is natural to it, a sense of increased warmth is felt. The surface, which is giving off more heat because more warm blood
is passing through it, not only feels warmer but is actually
warmer than it was before. Something else also happens
which will be more fully described in another lesson, but must
not be altogether omitted here. When this effect of alcohol
on the blood-vessels is set up, the rate of the passage of the
blood through them is changed because the heart beats with
increased rapidity. Under this there is warmth developed by
motion, and that warmth is given out also from the surface
and increases the sense and feeling of warmth, like the glow
produced by rubbing or friction which causes surface heat.

The warmth of surface which seems to be imparted by alcohol only *seems* then to be imparted. Positively the warmth
is not imparted by the alcohol, but is set free by it. The

warmth that is felt is given off at the expense of the body, which is making it by the chemical changes going on for its production, and it is escaping from the surface owing to the weakness of the vessels, and that greater exposure of surface which has been caused by the alcohol.

When my hand was suffering from hot-ache, induced by exposure to intense cold, the vessels, weakened by the cold, became unable to resist the blood when the constriction from the cold was taken off, and so the redness was seen and the sense of heat was felt as the blood recharged the vessels. Alcohol acts in the same manner as the cold.

Questions on Lesson XXVI.

1. Is the action of alcohol in causing a first stage of effect common in all classes of animals?

2. Give an illustration of a sense of heat of part of the body produced without the application of heat.

3. In the case given in illustration, why is a sense of warmth produced?

4. What effect has alcohol on the minute blood-vessels of the surfaces of the body, and of the course of the blood through them?

5. Why is the sense of heat felt after partaking of alcohol?

6. Why do cold and alcohol act alike in producing the sense and feeling of external warmth or heat?

LESSON XXVII.

ANIMAL LIFE UNDER ALCOHOL.—THE SECOND STAGE.

THE effect of alcohol on the warmth of the body is more distinctly traced as the after stages of its influence are developed. So far I have noticed and explained the first stage alone, that stage of excitement, when the face of the drinker is flushed, and his manner more or less excited. In this stage the man is often said to be "jolly," and what is commonly called the jollity of drink is felt by him. It is not a very

happy affair after all, if what has to be paid for it be measured
by the brevity of the pleasure. For, unfortunately, there are
second, and third, and fourth stages of the action of alcohol
on the body, which are anything but happy stages.

If the quantity of alcohol taken be comparatively moderate,
if it be but sufficient to excite without actually causing intoxication or drunkenness, there comes on a second condition
or stage (assuming even that no more alcohol be taken), during
which the flush of the face and skin dies away, the mind
begins to get rather bored and languid, and there is felt a
slight chilliness of the body. When the thermometer is now
brought into play, and the temperature of the body is carefully
taken, we find that the animal warmth is falling.

It does not fall very low, however, if the person is in a
very warm room, and has been provided lately, or is provided,
with sufficient food; but the fall is marked, reaching often,
under ordinary circumstances, to a degree on Fahrenheit's
scale. Should the person in this stage go out into a cold air,
and especially should he go out into a cold air while he is
badly supplied with food, the fall of temperature will be
much more decisive. It may reach a decline of two degrees
below the natural standard. To the person himself the condition is almost painful: he is depressed and chilled. In this
state he easily takes cold, and in frosty weather readily con-
tracts congestion of the lungs, and that disease which is known
as bronchitis. Nothing is more common in winter time than
the production of disease from this cause. When I say that
in our country alone thousands of persons are affected, in the
manner described, during sudden changes of season, from warm
or mild to cold weather, I do not at all over-estimate the
danger.

It takes some time for the body affected by alcohol to regain its natural warmth. Under very favorable conditions it
may become warm again to the proper standard in two hours.
Under less favoring conditions three hours may be wanted,
and a longer period still in case the person affected be exposed
to severe cold, imperfect supply of food, or fatigue, all of
which occurrences often happen in combination.

Many a man who has been busy all day and has not had a
really good meal of hot or nutritious food, rushes off when his
day's work is done to some inn or house, a long way from his

own home, to join a few convivial friends, and, as he says, to make himself merry. He sits in a warm room ; he relaxes his blood-vessels with some strong drink ; he smokes his pipe ; he laughs and talks, and perhaps sings, until he is tired ; and then, exhausted in this manner, without being more than elated, not tipsy at all nor anything like it, he starts off home through the cold air. In a few minutes the cold snatches from him his warmth so fast he feels as if he had taken too much drink ; he almost reels. He gets home weak, excessively cold, shivering, and perhaps bedewed all over his body with moisture from his own perspiration, condensed like dew on his chilled body. Next morning he is feeble for his work, at the best, and it is fortunate if he is not laid up with cold at the chest or with rheumatism. Who can wonder at it ?

Questions on Lesson XXVII.

1. How many more stages of alcoholic action are there after the first stage of excitement ?

2. How does the warmth of the body change in the second stage ?

3. What sensations are felt in the second stage ?

4. What length of time is required for the body to come back to its natural state from the second stage ?

5. How do men endanger themselves during the second stage of alcoholic action ?

6. What diseases are specially apt to occur from this exposure to danger ?

LESSON XXVIII.

ANIMAL LIFE UNDER ALCOHOL.—THE THIRD STAGE.

A MAN who has partaken freely of strong drink, but has not taken enough to make him intoxicated, may recover from that second stage of alcoholic influence without showing any signs of suffering beyond what were described in the last lesson. If, however, he should imbibe sufficient to affect him to intoxication, a third stage of action is reached, and is very serious. In

this stage the great vital organs of the body, the brain, the
lungs, the liver, the kidneys, are all too full of blood, and in
the most unfit condition for the performance of their work.
The nervous system, through which all the acts of the body
are directed—the movements of the body, the thinking, and,
in fact, all the working powers—is specially deranged. The
brain is obscured, and the mere animal or passionate nature of
the man is allowed full play, uncontrolled by the reason and
judgment. Men, therefore, in this stage are seen in their most
ridiculous tempers. Some men are horribly passionate, violent,
and cruel ; others are silly and talkative, telling sometimes
things of themselves they ought to be ashamed to hear, and of
which it would be prudent for them not to speak ; or laughing
insanely at sayings which are not clever ; or boasting, or utter-
ing untruths ; or crying and bewailing, as trials too hard to be
borne, commonplace griefs which sober men would think it a
waste of time to name. During this third stage of alcoholic
influence the man is unsteady in his movements ; he cannot
direct his muscles as he would ; he cannot put his hands steadily on the things he wishes to reach, and when he tries to walk
his gait is infirm or reeling.

At this time he is indeed a mere wreck of a man in mind
and in body—an object of pity and often of ridicule. He is so
weak that a child, if it had the courage, could topple him over ;
and he is so childish that clever, cunning knaves can impose on
him. What is worse than all, he is, for the moment, in a state
of disease throughout the whole of his vital organs. These
organs may recover sufficiently, and do recover so far as to receive, at first, no very evident harm ; but, in truth, from the
first some little harm is done, and after the evil has been inflicted a few times the results are so severe they are never entirely recovered from. The blood-vessels are subjected to
extreme strain, they lose their elasticity, and the blood circulates through them with difficulty and with irregular motion.

In the third stage of the action of alcohol the motion of the
heart begins to get feeble, and the heat of the body soon becomes greatly reduced. The thermometer will now sometimes
indicate that the animal warmth is reduced as many as one, or
one and a half degrees Fahrenheit. But now and then in this
stage there is a short rise of temperature, if the drinker, feeling
his exhaustion to be great, is induced, as he too often is in-

duced, to take another extra glass of stimulant, in order to sustain, as he foolishly supposes, his failing powers.

In the end, in the common run of cases, the third stage of alcoholic action, marked as it is by coldness of body, weakness of body, weakness of mind, and uncertainty of movement, ends in listlessness or sleep, during which recovery from all the injuries that have been inflicted is exceedingly slow. The warmth of the body does not return in its full degree for a very long period. When the affected person first wakes from sleep the warmth may rise a little, but it soon goes down again. Under a new supply of alcohol it may rise a little, but it soon goes down again, and a whole day may pass before it is once more steadily preserved at its natural standard.

Questions on Lesson XXVIII.

1. What is the condition of the important organs of the body during the third stage of the action of alcohol ?

2. What foolish acts do men commonly perform during this stage ?

3. How is the motion of the muscles of the body affected in this stage ?

4. How is the motion of the heart affected during this stage ?

5. How is the animal warmth affected during this stage ?

6. How long a time elapses after this stage before the warmth of the body is restored to its natural standard ?

LESSON XXIX.

ANIMAL LIFE UNDER ALCOHOL.—THE FOURTH STAGE.

THERE is still another stage during the action of alcohol which I have named as the fourth stage. It only precedes that final stage of all, death, to which some reach while they are under alcohol, although the event is rare.

When a man has arrived at the fourth stage, it is said of him, in rude but expressive words, that he is "dead drunk." The near approach to actual death in which the victim of drink now lies, is completely expressed by this phrase. He is not

dead, but he is dead drunk.   He is next door to dead.   He is
dead to the world, for he can neither hear, nor see, nor feel.
His limbs, like the limbs of a dead man, drop down helpless
when you raise them.   He is not quite so cold as a corpse, but
he is so cold the touching of him reminds you, with a shudder,
of a something that is corpse-like.   He is indeed at the gate of
death, and but for the gasping, rattling, heavy breathing, with
now and then a deep snore, the unskilled looker-on would
think he was dead.   It happens sometimes actually that a
doctor has to be called to men in this condition, in order to
determine by skilled knowledge of the signs of life, whether
life is or is not extinct.

I think there is no more awful spectacle for any one to see
than that of an unfortunate man or woman brought, in this
manner, to the edge of the grave by their own act and deed.
It were well if all young people would shrink from the thought
of entering into such a condition as they would from the
thought of sinking into deep waters to drown there.

During this fourth stage of alcoholic influence all the active
life of the body is in abeyance except one part, and that fortunately is a part which is not left under the self-control of
man.   If this part were not removed from the control of man,
every one who is dead drunk would die.   But this part lives.
It is the centre of the nervous system which governs the movements of the breathing.   When all other living parts fail,
when the brain is obscured, and when the muscles, which we
move by our will, are powerless, this part lives ; it keeps alive
the breathing, and the breathing suffices to keep the heart
moving towards recovery.

How seriously even the vital centre of breathing is endangered, the bad, irregular, and noisy breathing of the drunken
man visibly and audibly testifies.   When a skilled observer
feels the pulse and listens to the beat of the heart of a drunken
man he may be surprised that the circulation of the blood does
not stop altogether, for often it seems to cease, and the muscles
of the body of the prostrate victim tremble and start as if the
act of death were being carried out.

I have said that the body is cold to the touch during this
stage of extreme action of alcohol.   I should add that when
the temperature or warmth of the body is measured by the
thermometer during this stage, it is found to be from two and

a half to three degrees below the natural standard. It is no wonder therefore that the body feels cold, and certainly it is no wonder that persons, when they are exposed in this state to severe external cold or to cold and to damp, either die on the spot, or recover from the immediate danger to find themselves afflicted with some internal disease, such as congestion of the lungs, bronchitis, or rheumatic fever.

It takes from two to three days, under the best of circumstances, for the animal warmth to become steadily re-established after the occurrence of this stage. Usually as the reaction from the cold comes on, there is a fever of the whole body, just as there is a sense of "hot-ache" when the hand which has been in snow begins to recover its warmth. In a word, the whole body is subjected to the extremest strain of its powers, and is thrown so completely out of gear for the time that I really doubt if a man who has once been through the fourth, or dead drunk stage of alcohol, is ever quite the same healthy man that he was before.

Questions on Lesson XXIX.

1. What further stage may occur from the action of alcohol on the human body ?

2. What are the signs of life in this stage ?

3. Why does the person usually continue to live through this stage ?

4. What change in the warmth or temperature of the body is observed in this stage ?

5. What diseases are often started from this stage of the action of alcohol on the body ?

6. How long a time is required for the complete restoration of the warmth of the body to the natural standard from this stage'

LESSON XXX.

THE STAGES OF ACTION OF ALCOHOL.

THE four stages of action wrought on the body by alcohol are now fully before us. It is well to observe that during every one of these stages something has happened to the body which is not in the order of Nature. The flush and increased warmth, or rather the sense of warmth, in the first stage is not natural, for it partakes for the moment of the character of a fever. The excitement of the mind during the same stage is not natural; it is exhaustive of the bodily powers, and exhaustive in most instances for no useful purpose whatever.

The chilliness and slight depression of the second stage is not natural. It shows that the body has been exhausted, and that now it is tired, as if it had been subjected to some extreme physical strain, which in truth it has. The want of perfect control over words and thoughts in this stage is not natural, for it shows the mind to be exhausted as well as the body.

The deficiency of muscular control of the limbs, the reeling gait, the thick speech, the confused thoughts, and the passion of the third stage of alcohol, are all signs indicating a diseased, as distinct from a healthy, state of body. The coldness of the body during the same stage cannot be natural; and the weakness and wandering of the mind must of necessity be considered as showing the most serious derangement of the mental functions and powers.

The last stage of all, the stage just short of death, the fourth stage of the action of alcohol is clearly not only unnatural, but a stage of dreadful disorder and of danger. It is clear, surely, to the simplest mind, to the mind of the youngest child who can read this book, that a person who is lying down unable to move naturally, unable to hear plainly, unable to see correctly, unable to speak distinctly, and unable to do any thing more than breathe and live—it is clear, I repeat, to the simplest mind that a person so placed must be in a state of danger and disease as bad as any that could be caused by those accidents which we all shrink from, accidents that wound, and stun, and kill.

If we look at the whole course of the action of alcohol from the first stage to the last, we can see no good whatever that is supplied by it. Every step that seems harmless is, at best, nonsensical ; and every step that seems to be hurtful, is hurtful beyond anything that I can explain in this little work.

But some people who have not thought carefully on this subject say, and argue very keenly while they are saying it, that strong drink does good when it puts people in good spirits, and makes them feel warm and comfortable. They admit that when the drinking is carried on to do more than cheer up or exhilarate, and make the body feel warm, it is a very injurious thing indeed. They admit that when the strong drink makes the limbs unsteady, the memory confused, the thoughts hazy, and the body cold, it does a great deal of harm for which they, even they, are exceedingly sorry. They admit all these facts, but they hold that drink must do good when it cheers and warms. To reduce their argument to a very plain matter of fact, they contend that alcohol does no harm so long as its action is confined to the production of the symptoms of the first stage of its action. We have seen, however, that this first stage is really a stage of reduced power ; that the warmth or glow which is felt, instead of being a real supply of warmth to the body, is a di-engagement or giving off of heat from the body ; that the cheeriness which is felt is not due to any increased strength of the body or mind, but to the quicker motion of the blood and to the actual rapid waste of the powers of the body and of the structures of the brain, through which those powers are displayed.

Thus the defence of alcohol has for its foundation the argument that it is good to seek temporary gratification from an agent which, by its action in this very direction, leads to evil and even fatal consequences.

Questions on Lesson XXX.

1. In what respects are the first and second stages of alcoholic action hurtful?

2. In what respects are the third and fourth stages unnatural and hurtful?

3. Why, during every part of these stages, is some harm being done to the body ?

4. What arguments are used to prove that alcohol is useful in some of these stages ?

5. In which stage is it held to be useful ?

6. What evidences tell against this argument ?

LESSON XXXI.

ALCOHOL AND COLD.

It will be guessed, from what has been said respecting the action of alcohol in setting free the heat of the living body from the surfaces of the body, that, in fact, the glow felt indicates the loss of heat, and means a process of cooling instead of warming of the body as a whole.   This view is quite true, and it accords with and reasonably explains all that is known of the after-effects of alcohol.   It explains clearly why men who are exposed to extreme cold suffer so severely from taking alcohol.   These men feel for a few moments warmed by the spirit ; they feel a glow, which lasts for a short time and is very pleasant ; but soon the keen cold air seizes their warmth, and snatches the pleasant glow away.   As nothing has been supplied by the alcohol to keep up the supply of heat, the vital energy is rapidly exhausted, and the exhaustion from cold becomes extreme, sometimes fatal.

In my inquiries I put to the test the action of cold with and without alcohol, on the same kinds of animals under precisely the same conditions.   I found, thereupon, that if two animals of the same kind were let sleep in a very cold atmosphere, with no other difference between them than that one went to sleep under the influence of alcohol and the other free from that influence, the one under the alcohol would sleep not to wake again, while the other would wake from its sleep in the natural way at the natural time.

Precisely the same fact has been observed in man, but until the reasons of it were made clear by the results of experimental inquiry, they were not understood as they now are.

The facts we have now learned teach us that alcohol, instead of keeping out the cold, allows the cold to rob the body of its heat. This is important knowledge, because it corrects an error which, I am bound to say, had an easy origin and a very plausible support. It was quite easy for a man to say that a something which made him feel warm for a few moments warmed him ; and it was not so easy or natural for him to detect that the after-excess of cold was due to that which he felt to be a warming agency. I hope we have at last got to understand the error, and the delusive argument on which it has so long rested.

The facts have yet another meaning, which deserves to be understood. There have been some persons who have taught that alcohol burns in the human body as if it were fuel, as if it burned in the body as it does in a lamp, giving out heat, in a form of slow burning. There is no doubt that the heat of the body is maintained by the combustion of substances which are capable of slow combustion, or *eremacausis*, as the late Baron Liebig, the great chemist of this century, called the process. There is no doubt that the substances which we take as food for the purpose of being burned in the body, such substances as sugars and fats, are of the same elementary form for combustion as alcohol, that is to say, are bodies containing hydrogen and carbon as their chief elements. It was not unreasonable, therefore, to conclude that alcohol would play the same part in the body, and should be ranked as a fuel food. That it does not play this part is, however, shown by the facts I have described. The value of a true fuel food is shown by its efficiency in keeping up the animal warmth, just as the value of a true fuel out of the body is shown by its efficiency in keeping up the heat of the fire. If alcohol, therefore, were a true fuel food, it would enable the body to resist cold instead of making it colder ; and in the extreme degrees of cold it would go on burning like other fuel foods, and would maintain, instead of helping to destroy, life.

But as alcohol checks the production of animal heat, so the mischiefs produced by it on the body are greatly increased by cold. Alcohol and cold go hand in hand in their action. The worst influences of alcohol on man are, therefore, observed when the weather is intensely cold. All great consumers of alcohol are chillier during the winter months than are those

who abstain, and as they who drink alcohol constantly labor
under the delusion that if they are cold they must take wine,
spirit or ale to keep them warm, they make matters still worse
by actually going to their enemy for relief.

Questions on Lesson XXXI.

1. What does the external rise of animal warmth which is found after
the taking of alcohol indicate?

2. Why do men who are exposed to cold after taking alcohol feel it
so much more intensely?

3. What facts prove that alcohol assists cold in enfeebling or destroying life?

4. Why has alcohol been supposed to act as a food useful for the
support of animal heat?

5. In what respect does it resemble foods that act as fuel foods?

6. How do the facts relating to the action of alcohol refute the view
that alcohol is a fuel food?

7. How do alcohol and cold act in relation to each other?

8. What grand mistake do such persons make in trying to obtain
warmth?

9. In cold weather, what is the effect on the body of moderate quantities of alcoholic drinks?

LESSON XXXII.

THE HEART UNDER ALCOHOL.

WHEN the hand is placed on the chest to the left side,
"over the heart," as we are wont to say, we feel the beating
of the heart. When we place a finger on the wrist we feel
a beat which we call a pulse, and if we put the finger in the
same way on the side of the throat, under the collar-bone, on
the temples, or on many other parts of the body, we feel there
a similar beat which we also call a pulse. If the beat of the
heart and the beat of the pulse be counted at the same time,
it will be found that they go together, stroke by stroke; and
the reason of this is that the beat of the pulse is caused by

the passage of blood through a blood-vessel, an artery, which proceeds from the heart, and which is filled with blood by the stroke of the heart. The heart, by its motion, pumps the blood all over the body through the arteries : the arteries are elastic tubes, and each pumping-stroke of the heart gives an impulse felt in the artery : this is the pulse.

In a grown-up man the heart beats, and the stroke of the heart and of each pulse-beat is felt, seventy-three times in a minute. This may be taken as a fair average of the number of strokes during the whole space of the waking hours. Seventy-three strokes a minute represents 4,380 strokes an hour, and 105,120 strokes in the twenty-four hours. But the heart beats slower when the body is lying down, and a little slower when it is sitting than when it is standing ; and we may fairly take off 5,120 strokes, and say that the number of beats of the heart each twenty-four hours is 100,000. By the work that is done in this way, over five thousand ounces of blood are pumped over the body by the heart in twenty-four hours. The value of this work is represented by saying that a weight of over 115 tons has been raised one foot.

We have thus before us the natural work performed by the heart of a healthy, full-grown man, under natural circumstances of food and labor, and without the interference of an alcoholic stimulant. Suppose, however, we make an interference by allowing the man to take into his blood some measure of alcohol, what then would happen ? Suppose one could see the pulse of the man beating when it is doing its natural work just as we can see the seconds-hand of a clock or watch making its usual round per minute ; and suppose again that, while watching, the alcohol could be brought into play. What then should we see ?

We should see many small changes of motions if we looked very closely and learnedly, but of these changes I need not stop to write. That which would strike us most would be : first, the fact that the beats of the heart became ever so much quicker than they were during the natural state ; and secondly, the further fact that the rapidity of the beats increased with the quantity of the alcohol that was swallowed. If the quantity of alcohol taken were four fluid ounces in the twenty-four hours, the number of beats of the heart would be increased, in that time, from 100,000 to 112,226. That is to say : 12,226 extra beats would be delivered in the time named, or a little over 509 extra strokes an hour, or very nearly eight and a half strokes per minute, beyond the natural number.

If the quantity of alcohol taken were six ounces by measure in the twenty-four hours, the number of beats of the heart would be increased in that time from 100,000 to a little over 117,388. That is to say : 17,388 beats would be delivered in the twenty-four hours, or 724 extra strokes per hour, or 12 per minute, beyond the natural number.

If the quantity of alcohol taken were eight fluid ounces in the twenty-four hours, the number of beats would be increased in that time from 100,000 to 124,045 extra. That is to say : 24,045 beats would be delivered in the twenty-four hours, or over 1,000 beats per hour, or nearly 17 per minute beyond what is natural.

Questions on Lesson XXXII.

1. What is the cause of the beat of the pulse ?

2. In a healthy man, how many beats of the heart and pulse are felt per minute?

3. What is the number of the beats of the heart and pulse in twenty-four hours?

4. How many extra beats of the heart follow the taking of four fluid ounces of alcohol?

5. What extra number of beats follow the taking of six ounces?

6. What extra number of beats follow the taking of eight ounces?

HEART-WORK UNDER ALCOHOL.

IT will be said by some who, in spite of all arguments against the use of strong drink, like it still, and defend its use ; it will be said by some of these that to take the quantity of alcohol named in the last lesson is to take too large a quantity. They will insist that if smaller or, as they would express it, moderate quantities be taken, no harm will result, and that the beat of the heart will not be too severely increased. It will be good for us to consider this point for a moment, because it never is wise to let an opponent go without a proper answer, and it is always wise, when an opponent gives a fair argument, to be ready to answer him thoughtfully from the best, most correct, and simplest evidence. Here then is an answer to the argument now before us.

We will suppose a very moderate drinker indeed. We will suppose a man or woman who takes three half-pints of ale a day : he or she would, as a working man or woman, be called a very temperate, sober person. We will suppose a man or woman who can afford to take wine instead of beer, and who takes two wine-glassfuls of such a wine as sherry at luncheon, and three at dinner : he or she would certainly be called extremely temperate. We will suppose another person who takes spirits instead of wine or beer, and who indulges in a wine-glassful of, say, whiskey mixed with water at luncheon, and the same at dinner, and with one glass more with water in the way of what is said to be a "night-cap" before going to bed. Such persons would pass in the world as most moderate and temperate in respect to drink. In fact all these people, beer-drinkers, wine-drinkers, spirit-drinkers, would in circles of alcohol-drinkers be considered exemplary, or perhaps would be notorious as persons who could not be induced to take as much as would do them harm. Let us see the results. When we calculate up the amount of alcohol which these very moderate people have swallowed in the twenty-four hours, we find that it amounts to two ounces of alcohol at least in each case. Two ounces of alcohol will, I find, raise the beats of the heart to 6,000 extra beats in the twenty-four hours, which means an amount of work represented by the act of lifting a weight of seven tons one foot high. Now let us see further what this means. One ton divided into ounces would give 35,840 ounces; so that the work done is really represented by the process of lifting a seven-ounce weight 35,840 times the height of one foot each time. Imagine that some one was obliged in twenty-four hours to lift so light a thing as a seven-ounce weight with one hand from a table, and put the weight upon a shelf one foot higher than the table. It looks a very little thing to do, and it is a very little thing to do a few hundred times, but if it has to be done 35,840 times in twenty-four hours, or 1,493 times an hour, the labor would be so great that the hand would fail in a few hours altogether. If in writing two or three hours, the inkstand be placed one foot above the table, the mere matter of raising the pen that one foot three or four times a minute becomes too fatiguing to be borne. How then must the heart be wearied when it is driven to the extra and unnecessary work of lifting seven ounces one foot high 1,493 times an hour. If a man were obliged to drive his heart to perform so much labor by running or other severe work, he would think his fate hard indeed, he would say it was like working at the treadmill or other similar laborious task, but he would not be more wearied, and he would not be so much injured.

Questions on Lesson XXXIII.

1. What argument is used by some persons who are in favor of alcohol to express that alcohol need not do injury to the heart?

2. How is this argument best met and disproved?

3. What quantities of different drinks are supposed by most people to be moderate allowances?

4. State the amount of alcohol that is conveyed into the body in what are held to be moderate allowances of alcoholic drinks?

5. What amount of extra work would such assumed moderate allowances of alcohol put upon the heart?

6. How would you illustrate that this would be a large excess of work for the heart to perform?

LESSON XXXIV.

ALCOHOLIC FATIGUE.

THE reason why those who are habituated to the use of alcohol believe that moderate quantities of it inflict no harm is, that they do not clearly know what it is to be free from its influence. A man who never indulges in alcohol has quite a different experience if, by accident, or for sake of the experiment, he partakes even of one fluid ounce of alcohol. I have seen the effect of this quantity on a youth who was not accustomed to take any alcohol at all, and on him the effect was extremely well marked. In a few minutes his face became very red, his pulse rose fifteen beats in the minute, his temperature rose half a degree Fahrenheit, and he complained that he felt a sharp throbbing in his head and a strange fulness. In twenty minutes these symptoms began to pass away, and in an hour they were all gone; but now his temperature fell nearly half a degree below the standard, and did not return to the natural state until after a light meal, two hours later. Similar phenomena may also be seen in the inferior animals which are unaccustomed to the taking of alcohol when they are subjected to proportionately small doses.

In like manner a man who is very temperate, but who takes alcohol, feels most distinctly the effect of even a slight excess. Such a man, if he be tempted to move from the single glass of mild dinner ale a day to take a glass or two of wine when he goes out to dinner, or to take a single glass of grog at night, is conscious of the evil influence the next day. He says, if he speaks truly, that he was rather excited by the drink; and he says, when the stage of depression comes on, that he feels "all-overish, depressed, rather chilly, and not up to the mark." He is tired, and he thinks he should be none the worse if he took an extra glass of ale to set him right. In nine cases out of ten he does take this extra glass of ale; it does set him what he calls right, and finding how easy a thing it is to get over a slight excess, the next time he is tempted he ventures a little further. So the habit of taking too much begins in taking

77

just a little, while being indeed very temperate, and while keeping in fashion with other folks.

This is the beginning of woe. If men could have been kept sober by practising temperance the fight against intemperance would never have had to be fought, such books as this I am now writing would never have had to be written, and such a fact as drunkenness would be looked on as a rare calamity. For let it be most solemnly impressed on the mind that the depression which the temperate man felt, because in an hour of weakness or folly he took one extra glass, was due to the failure of his heart. That hard-worked organ, which in sleeping and waking knows no rest, had been overworked. Probably at the close of a day of hard work, when the heart wanted the body, which it had fed all the day through, to lie down and rest, with the blood running on the easy level plane, probably, at that critical time, the owner of the body kept himself up unusually late, and at the same moment, by the alcohol he took, whipped on the wearied heart to beat some five or six thousand extra strokes, and to lift six or seven extra tons weight.

Is it wonderful that the heart should get weary, and should tell its tale of weariness to the whole body, and crave for more rest the next day? I think not. If other causes of weariness, if having to walk for half the night, or race, or row, or lift weights, or ring bells, or any other excess of work wearied the heart, no one would wonder, or want to do anything more than lie down and rest. But when alcohol does this wrong, the fatal remedy tried is more alcohol.

Questions on Lesson XXXIV.

1. Why do people who indulge in alcohol so often fail to recognize its bad effects?

2. What effects are produced by a small quantity of alcohol on those who have never taken it?

3. What sensation does the temperate man feel if he slightly exceed his usual quantity of alcohol?

4. How do persons who take alcohol try to remedy the mischief that has been induced by it, and to what danger does the plan lead?

5. Why is so-called temperance without abstinence an insufficient safeguard against danger of excess?

6. What vital organ of the body fails first in power after indulgence in strong drink?

LESSON XXXV.

ALCOHOL AND MUSCULAR STRENGTH.

THE effects produced on the heart by alcohol are not confined to the weariness which is at once caused. The heart itself is very soon deranged and enfeebled for its work, if alcohol be long taken as drink. The heart is a muscle, and it moves by the same means as other muscles do in order to carry out its duties of driving the blood all over the body. It may be well, therefore, to touch on the subject of the action of alcohol on the muscles generally, as well as on the heart, which is also a muscle.

The muscles of the body are those masses of flesh which *cover* the bones of the limbs and trunk of the body, and which are enclosed within the covering called the skin. They are like little engines, and by them the body is moved, and the limbs are moved. If you feel the fore part of your arm above the elbow while the arm is straight out, the muscle is felt to be relaxed and quiet. If you bend your arm on the elbow-joint, the muscle is felt to be round, and prominent, and firm. The muscle has contracted, and, in the act of contraction, has lifted the hand and arm upwards. The heart is a hollow muscle made up of four muscular cavities, and when the walls of these cavities contract, they propel the blood onwards. The muscles are made to contract under what is called the stimulus derived from the nerves. In some instances the muscles receive the stimulus to act from the will. In other instances the stimulus is supplied without the will. That muscle which moved the arm is under the control of the will. The heart is not under the control of the will; if it were, we might sometimes, when we are very busy or anxious, forget that the heart had to be kept moving, and then it would stop, and we should faint or die.

Such are some facts about the muscles of the body. I could tell many more, but these are sufficient for my proper subject—the action of alcohol upon the muscles.

Alcohol then acts on these muscular engines in a special manner. It is often thought that wine, and beer, and spirits give strength to a man, that they make the muscles contract with more force, and sustain the action. I have put this matter to the test by means of experiment, and I have found that the idea of alcohol giving force and activity to the muscles is entirely false. I found that alcohol weakens the muscular contraction, and lessens the time during which the contraction can continue active.

In the first stage of the influence of alcohol, enfeeblement of muscular power commences, and it continues into the other stages with increasing effect, until in the third and fourth stages it is lost almost altogether. This is one of the reasons why the man who is suffering from much alcohol is so utterly prostrate and helpless. It has been found also by other observers that men who have a great deal of work to perform of a muscular kind, men who have to march as soldiers, to walk long distances against time, or to row with great force and rapidity—as boat crews do—that all such men carry out their work much better when they avoid every drink containing alcohol.

If two men of equal age, skill, and build were to contend in a feat of strength, and one of them indulged in alcohol and the other did not, the chances of success would be all on the side of the abstainer, other things being equal. Through every muscle of the alcohol-drinker that would be called into action by the contest, the bad influence of the alcohol would be felt, but through none so much as the great central muscle on which all the rest depend—the heart.

Questions on Lesson XXXV.

1. What is the heart as a living structure of the body ?

2. What are the two classes of muscles of the body, and what are their duties ?

3. From whence is the stimulus which excites the muscles to contract derived ?

4. What is the common idea as to the value of alcohol for giving strength to the muscles ?

5. Is this idea founded on any known fact, and is it true or untrue ?

6. What is the actual effect of alcohol on muscular power ? Give some proofs of its action in this respect.

LESSON XXXVI.

ALCOHOL AND THE SMALL BLOOD-VESSELS.

CONNECTED with the influence of alcohol on the muscular system, we have to consider its action on the minute blood-vessels of the body. I have stated that with each stroke of the heart the red arterial blood is driven through the tubes called arteries all over the body, in order that it may nourish the body in every structure of it, and be the means of permitting the body to be supplied with its vital warmth. These vessels or arteries which convey the blood are at first large tubes, but they soon commence to divide into smaller tubes, and at last they divide into very minute tubes indeed, which are invisible to the naked eye, and are spread out like a fine hair-like network through the different organs which compose the body. Fine as these tubes are, they let the blood pass through them; but they do not let it pass in such a manner as to be under no control. Fine as they are, they retain the power of contracting on the blood, and of precisely regulating the due supply of blood that shall go to every part.

The regulating control of these small arterial vessels is not under our will, but under the involuntary nervous stimulus; and anything that greatly affects the nervous system interferes with the regulation. When some impression is made directly on the nervous system, or indirectly on the vessels by a mental process—as when something is said to a person which makes him feel ashamed or timid—the shock may cause the little vessels in the cheek to lose for a moment their nervous stimulus, and the cheek then becomes crimson, because the control of the vessels being lessened, the heart pumps a little more red blood into the vessels, and the crimson flush is produced. When the surface of the body is exposed to cold, the little vessels become weakened by the cold; and when the pressure caused by the cold is removed, the heart throws more blood into the vessels, and the tingling sense and redness which follows indicates the weakness that has been caused.

81

If the nervous shock or the cold be extreme, the vessels may lose all power of resistance, and the effect may extend to the heart itself, so that the circulating machinery, through its whole length, is weakened. Then the surface of the body becomes intensely pale, and all the body is for the time enfeebled. We say the person so affected, as from fright or passion, is pale or white with fear or rage, and we notice how feeble he is; it is remarked of him that he might be knocked down with a straw. Or we say, if a man is brought to the same state by cold, that he is stricken by cold, and looks like death.

These are the effects of nervous shocks and exposures to cold; but the same effects may be produced by other means. Some substances taken into the body by the mouth produce just the same results. They weaken, or to speak technically, paralyze the controlling blood-vessels. Many agents do this, but not one more potently than alcohol. The grand physical action of alcohol on the body is that of weakening the minute blood-vessels that regulate the supply of blood to the body at large.

Questions on Lesson XXXVI.

1. What vessels convey the blood all over the body from the heart?

2. What duty do these vessels perform at the end of their course in respect to the passage of blood through them?

3. Under what regulating control are the vessels placed?

4. When the contractile power of these vessels is lessened, what is the result?

5. Give some illustrations of nervous derangement affecting the minute vessels.

6. What is the action of alcohol on the regulating power of the minute vessels?

LESSON XXXVII

ALCOHOL AS A STIMULANT.

To have a correct knowledge of the action of alcohol on the minute blood-vessels of the body, is to know a great deal as to the mode in which it produces its effects on the body altogether. In the first stage of alcoholic action, with which we are now conversant, when the face is flushed with blood, and all the surface of the body is in a glow, and when the mind is excited and the thoughts rapid and free—in this first stage the minute controlling vessels are relaxed and over-charged with blood. It is in this stage that the heart is beating so quickly. It is beating quickly because the resistance which held it in proper motion and check is taken off, so that it works away without being under proper regulation. If we take the pendulum from the clock, we know what happens; the clock begins rapidly to run down, because the main-spring which supplied the clock with motion is left to exhaust itself without check. It is just the same with the heart when the regulating check is removed.

During the first stage of alcoholic influence, it will be remembered that the body feels warmer than natural. It does so because the minute vessels on the surface are charged with a more than natural amount of warm blood, and from that there is a greater giving off or radiation of heat, and so there is felt the glow and sense of warmth, as if more heat were being produced. It is really not so; more heat is being lost than is natural, but no more is being produced. In this stage, indeed, the body is really cooling, and the coldness in the after-stages is due in part to the loss that is occurring in the first stage.

During this first stage the mind is exceedingly active. It is so because now a very free and unnatural volume of blood is passing through the brain. The vessels of the brain are weakened or relaxed just as are the vessels of the skin, and so the brain is excited to action, and for a time is very busy.

83

If you watch the company that makes up a large dinner-party at which the wine passes freely, you see that at first the persons at the table are all quiet and sedate, and that their faces are of a natural fresh color. But as the wine goes round and round, there is soon a change both to sight and hearing. The faces of the company become beaming and red, their eyes are very bright, and the motions of their hands and features are rapid compared with what they were at the first sitting down. The quiet, subdued conversation changes into quite a loud noise: the voices are louder and sharper, and the words quicker and quicker in delivery. What was a smile a few minutes ago becomes a laugh, probably a loud laugh, and so all goes on until there is a perfect din and clamour. It is now that the table is said to be in a roar.

There is presented here on a large scale the first stage of the action of alcohol. Everything that alcohol can by any means do usefully for the world is seen in this stage. This is its *summum bonum* or chief good. It was to enliven the feast after this fashion that wine first became fashionable in the history of man. The ancients rarely used wine for any other purpose than the one now specified.

To make the heart beat quicker, to make voices that were silent noisy, to make minds that were dull communicative, to make mirth flow freely, to enliven the countenance, to provoke laughter on very small provocation—these were the reasons for wine in primitive times. The reasons suggest wine as a luxury, but such a luxury, unfortunately, as the richest man cannot afford to enjoy.

Questions on Lessons *XXXVII*.

1. In what state are the minute blood-vessels which regulate the distribution of blood during the first stage of alcoholic action ?

2. Why does the heart beat so quickly under this state ?

3. Why does the body feel warmer than is natural under this state?

4. Why is the mind so much more active under this state?

5. What are the phenomena of the effects of wine during the first stage as they are seen at table when wine is taken freely ?

6. What view as to the primitive uses of wine do these phenomena suggest?

LESSON XXXVIII.

STIMULATION AND DEPRESSION.

The men of a past time called alcohol a stimulant, for which they are not to be blamed. They were quite correct in calling it so if the word be rightly understood. If it be meant by the term stimulant, that a stimulant is something which provokes action in a living body without supplying anything to fill up the expenditure of matter and force that is caused by the action, then alcohol is a stimulant. Just in the same way a blister is a stimulant to the part where it is applied, and a whip is a stimulant when it is used to make an animal move more quickly, and snuff is a stimulant when it makes one sneeze.

There are some stimulants which are much stronger than alcohol, but however strong they may be they give no strength to the body, they only excite it to action, and to wear out more quickly.

Therefore, during all the first stage of excitement from alcohol, when the heart is going so fast, when the breathing is so fast, when the mind is so active and readily moved, there is progressing, in fact, the most extensive and costly waste of bodily substance and power. Life is going on at express rate of motion ; all the glands of the body, by which the waste products are carried away, are being taxed to their utmost ; the heat of the body is passing off more rapidly than is natural ; the muscular movements are being carried out at the wasteful expense of animal power ; and the mental over-activity is a last, and, of all the rest, a most severe process of exhaustion.

These combined losses leave the body much fatigued, even when the process of alcoholic excitement is stopped at the first stage. When it is not stopped at the first stage, when under the hope of keeping up the excitement the stimulant is persisted in, the relaxation of the second, third, and, in extreme cases, the fourth stage is developed.

In these stages, as we already know, the skin is cold, the muscles are relaxed, the heart is feeble, the mind feeble or excitable, the whole body deranged and exhausted.

85

Every excitement is followed by exhaustion and need for rest, that the exhaustion may pass away.

When a man receives a severe shock from a blow or from some great mental fright or alarm, he may be paralyzed at once to the heart itself, and may fall faint, cold, and lifeless, or all but lifeless, to the earth. If he take, too rapidly, a large dose of alcohol he may fall in the same way. The first, second, and third stages of alcoholic action may be passed over, and he may fall prostrate and insensible at once.

In short, in whatever way alcohol acts on the body, whether it acts slowly and by successive stages, or rapidly, so as to produce all its evil action in one sharp charge, it acts as a reducer of the powers of life. Never let this lesson be forgotten in thinking of the effects of strong drink, that the drink is strong *only to destroy*; that it never by any possibility adds strength to those who take it, and that to resort to it for the sake of getting strength from it is like seeking for strength in exhausting and tiring exercises.

Alcohol, like these exercises, produces weariness, but the weariness is worse than that from any kind of labor. Good, muscular exercise quickens the motion of the blood and causes muscular exertion for a useful end. It enables the body to throw off its worn-out matter, and it invokes a good appetite for food, out of which new matter is introduced into the body to take the place of the old and used-up substances that have already played their part. But alcohol does not act like exercise in these respects, it excites to waste, leads the body at the same time with its bad self, checks the proper excretion of waste products, reduces the appetite, and impairs digestion.

Questions on Lesson XXXVIII.

1. What is the true meaning of a stimulant, and why is the term correctly applied to alcohol?

2. When alcohol is acting as a stimulant to the body, what harm is it doing?

3. What may be the immediate effect of taking a large quantity of alcohol?

4. What is the secondary action of all stimulants on the body?

5. What is the effect of moderate muscular exercise on the body?

6. Why is alcohol, which quickens action, hurtful, while exercise, which does the same, is useful?

LESSON XXXIX.

ALCOHOL AS A POISON.

THROUGH the influence exerted by alcohol on the minute blood-vessels which regulate the supply of blood to the different organs of the body, all the most vital and important organs are subjected to disease after its long-continued action. The features of those who indulge in strong drink exhibit the evidences of the evil committed by this subtle foe. The vessels, over and over again distended with blood, show at last the fact of such distention in those dark red, or all but blue, discolorations of the face, which are so often seen in the hard drinker.

These signs are but external proofs of deeper injuries of a similar kind. The vessels of the brain, the vessels of the lungs, the vessels of the heart, the vessels of the liver, the vessels of the kidneys, the vessels of the stomach, are likely to be in the same condition. The vessels generally are so weakened, they can bear very little pressure, and sometimes they give way, let out blood into the organs to which they belong, and endanger, suddenly, the life of their owner. Through these vessels the blood circulates more slowly than is natural, and so, under an extra cause of depression, the organs which are supplied by them become diseased and dangerously damaged. For these reasons, hard drinkers die in great numbers from congestion of the lungs during cold weather. The cold adds to the weakness of the vessels, and their failure is thereby made more complete. Again, in very hot weather, when under the heat the heart beats with greater rapidity and power, the weak vessels are apt to give way, with a not uncommon result of immediate death. Very often when it is said that people are killed by sunstroke, this is the mode of death. Some of the vessels of the brain within the cavity of the skull have become relaxed and weakened from alcohol, and during the intense heat, partly from the over-action of the heart, and partly from the expansion of the blood by the heat, the vessels have given way in the brain, and death has instantly occurred. Or, if de-

87

layed for a short time, death has taken place after a seizure of what is called stroke or apoplexy.

These are the results of a dangerous kind which come on in persons whose vessels are weakened by the use of drinks containing alcohol. They are not all the evils that follow from the same cause; nay, they are a small part of them. In all who are accustomed to take alcohol to a degree sufficient for them to feel an effect from it, some one or other of the organs of the body are sure to suffer from slowly-developed disease. The thin membranes in which the internal organs are enveloped become thickened and lose their elasticity; the structures of the organs themselves are made to undergo changes which render them unfit for their duties, and, to use a very homely but strictly correct phrase, the organs get to be old and worn-out before their time. Sometimes it is the heart that becomes weakened, relaxed, and unfit for its work; sometimes the lungs are rendered liable to a peculiar form of consumption; sometimes the liver becomes hard, shrunken, and so irregular in appearance that it looks "hob-nailed" on its surface; sometimes the brain becomes too firm in portions of its structure, or softened, or obstructed in its circulation; sometimes the kidneys undergo fatty or other changes which cause disease; sometimes the stomach becomes too weak to digest food. From one or more of these diseased changes death takes place.

Questions on Lesson XXXIX.

1. What are the effects of long-continued disturbance from alcohol on the minute blood-vessels?

2. What are some of the external and visible signs of these effects?

3. What other changes do the external signs often suggest?

4. To what special dangers are persons whose vessels have become weakened by alcohol subjected, in cold weather and in hot weather?

5. What slow changes are produced in the organs of the body by the drinking of alcohol?

6. What great organs of the body are affected by alcohol, and in what manner are they affected?

LESSON XL.

DISEASES OF THE BODY FROM ALCOHOL.

IT will easily be understood from the preceding lessons that alcohol is a cause of many diseases amongst those members of the human family who insist on drinking it, instead of trusting entirely to water as the natural beverage.

Alcohol produces many diseases; and it constantly happens that persons die of diseases which have their origin solely in the drinking of alcohol, while the cause itself is never for a moment suspected. A man may be considered by his friends and neighbors, as well as by himself, to be a sober and a temperate man. He may say quite truthfully that he never was tipsy in the whole course of his life; and yet it is quite possible that such a man may die of disease caused by the alcohol he has taken, and by no other cause whatever. This is one of the most dreadful evils of alcohol, that it kills insidiously, as if it were doing no harm, or as if it were doing good, while it is destroying life.

Another great evil of it is, that it assails so many different parts of the body. It hardly seems credible, at first sight, that the same agent can give rise to the many different kinds of diseases it does give rise to. In fact, the universality of its action has blinded even learned men as to its potency for destruction.

Step by step, however, we have now discovered that its modes of action are all very simple, and are all the same in character; and that the differences that have been and are seen in different persons under its influence are due mainly to the organs or organ which first give way under it. Thus, if the stomach gives way first, we say that the person has indigestion or dyspepsia, or failure of the stomach; if the brain gives way first, we say the person has paralysis, or apoplexy, or brain disease; if the liver gives way first, we say the man has liver disease, and so on.

Besides the diseases of the organs named there are others

89

that are favored by alcohol which are extremely painful to bear.   Gout is one of these diseases, rheumatism is another.

I must name one other disease specially, because it is so common.   I refer to derangement of the stomach, or indigestion.

All persons who indulge much in any form of alcoholic drink, are troubled with indigestion.   When they wake in the morning they find their mouth dry, their tongue loaded, and their appetite bad.   In course of time they become confirmed " dyspeptics," and as many of them find a temporary relief from the distress at the stomach and the deficient appetite from which they suffer by taking more stimulant, they increase the quantity taken and so make matters much worse.   They now become actually ill from weakness of stomach and imperfect feeding ; their breath becomes offensive, and soon the mind is depressed and languid.   Such persons, in very many instances, fall lower and lower into the vice of drinking heavily.   They feel as if they could not live without their fatal master.   They tell you it is both food and drink, and in this delusion they persist until they are made the victims of deadly disease from its use.

––––––

Questions on Lesson XL.

1. What is one of the worst and most deceitful evils arising from the use of alcohol ?

2. What is another of these deceitful evils ?

3. Give some illustrations of the mode in which alcohol produces disease.

4. What organ of the body is most commonly affected by alcohol?

5. What are the signs of this affection ?

6. What dangers do persons suffering from this affection specially encounter ?

LESSON XLI.

DEATH FROM ALCOHOL.

THERE are a great number of diseases caused by alcohol, some of which are known by terms that do not convey to the mind what really has been the cause of the diseases. They are : *(a)* Diseases of the brain and nervous system : indicated by such names as, apoplexy, epilepsy, paralysis, vertigo, softening of the brain, delirium tremens, dipsomania or inordinate craving for drink, loss of memory, and that general failure of the mental power, called dementia. *(b)* Diseases of the lungs: one form of consumption, congestion, and subsequent bronchitis. *(c)* Diseases of the heart : irregular beat, feebleness of the muscular walls, dilatation, disease of the valves. *(d)* Diseases of the blood : scurvy, excess of water or dropsy, separation of fibrine. *(e)* Diseases of the stomach : feebleness of the stomach and indigestion, flatulency, irritation, and sometimes inflammation. *(f)* Diseases of the bowels : relaxation or purging, irritation. *(g)* Diseases of the liver : congestion, hardening and shrinking, cirrhosis. *(h)* Diseases of the kidneys : change of structure into fatty or waxy-like condition and other changes leading to dropsy, or sometimes to fatal sleep. *(i)* Diseases of the muscles : fatty change in the muscles, by which they lose their power for proper active contraction. *(j)* Diseases of the membranes of the body : thickening and loss of elasticity, by which the parts wrapped up in the membrane are impaired for use, and premature decay is induced.

But it constantly happens that when deaths from these diseases are recorded, and alcohol has been the primary cause, some other cause is believed to have been at work.

To give some idea of the extent of disease and death which is caused by alcohol, I may narrate a striking fact relating to one class of men, the class which makes its living, and I had nearly said its dying also, by strong drinks. This is the class called publicans, hotel keepers, or beershop keepers. The publicans, as a class of men, seem very comfortably placed in the world. They live in nice warm houses, they have plenty of food on their tables; they are not obliged to go out in all weathers, and at all hours of the night, they are not exposed to great dangers like men who drive railway trains; they are not exposed to infectious diseases like doctors; they are not exposed to cold, and heat, and wet, and privation like agricultural laborers. In short, they are placed in more favorable circumstances than almost any other class of the community, except in this one respect, that, having alcohol always before them or near them, they are constantly tempted to partake of it. The effect of this practice is shown in their mortality. With all their other advantages to back them up, they die so much faster from disease caused by alcohol than the rest of the people, that in England, as the public records tell us, 138 publicans die in proportion to 100 of the whole of the community who are employed in seventy leading occupations. Even railway-drivers and servants, who are exposed to so many dangers, die at the lower rate of 121 to 138 publicans.

Questions on Lesson XLI.

1. What fatal diseases of the nervous system are induced by alcohol?
2. What fatal diseases of the lungs and heart are caused by alcohol?
3. What fatal diseases of the liver and kidneys are occasioned by it?
4. What diseases of the muscles and the membranes are caused by it?
5. What great error is often made as to the cause of these diseases?
6. What is the death-rate in England of the class of persons who sell strong drinks, as compared with the death-rate of people following other occupations?

LESSON XLII.

INSANITY FROM ALCOHOL.

I HAVE put forward the evidence that bears on the action of alcohol on the body. This evidence is very sad; but it is not all that has to be told. We have yet to learn what is the baneful influence of alcohol on the minds of men and women who indulge in its use.

There is no one who, if he be old enough to observe and think, has not been obliged to witness some of the humiliating scenes which are due to alcohol. He has seen some unfortunate man or woman under the influence of alcohol, intoxicated by it. In this strange scene he has had before him the evidence of the folly and wickedness to which strong drink is capable of subjecting women and men. He has seen how very silly the drunken man is: how angry he often is about trifles and little things; how he laughs, and raves, and cries, and says things which are not true, and imagines things which are not likely to be true; how he looks, and acts, and speaks, indeed, as if he were bereft of his senses. It is the fact that, for the time, he is bereft of his senses; he is a man who has gone mad. He spends his money madly, he treats his friends madly, he treats himself madly. Those who would love him best if he were not mad are now afraid of him, and often hide themselves from him; and well they may. For this man, in his madness, may hurt them, strike them, kill them; and when he has come to his senses, may find himself in prison, may be told by the gaoler that he is there to answer for the death of some one he has hurt or killed, may even have to be tried for his own life, and all the time he may remember nothing of that which has occurred to bring him, and which has brought him, to his awful condition.

This is madness from drink in men or women who are not yet permanently mad.

You who now know how alcohol influences and changes the brain, how it weakens the vessels of the brain, and why it

93

does so, will not wonder at all that such madness should follow upon the drinking of strong drink.

Neither will you wonder now to hear how often these temporary outbursts of insanity or madness from strong drink pass into permanent madness. We have traced how diseases caused by alcohol in other organs of the body than the brain, progress until they kill the body ; and we are prepared to understand that the disease of the brain, caused also by alcohol, progresses, when it has once set up, until it kills the mind as well·as the body.

But how far this disease of the mind extends amongst mankind no one yet has been able to tell ; for it so happens that the madness from alcohol is not confined to the one person in whom it first was developed. The madness may, and sometimes does, extend to that person's children, and becomes what is known as an inherited or family disease.

In order to combat the insanity to which men and women are subject, we have to build large places which are called asylums for the insane. Here the poor mad people are kept together and watched, that they may not injure themselves or others, and that from their mad ways there may be no danger. Amongst those who are made to enter these asylums, many are brought there in consequence of their having been made mad by alcohol, and it is certain that no other single cause of madness is so frequent. In one asylum it has been found that forty out of every hundred persons who were admitted insane had become insane, directly or indirectly, from the effects of strong drink. There are some of these who sit all day as if they were helplessly under the effects of alcohol. These have nearly all come into that state by alcohol, and are said to be suffering from general paralysis. This living death is one of the many terrible evils which alcohol inflicts on an erring and a foolish world.

Questions on Lessson XLII.

1. What conditions of mind does alcohol cause when it is taken in large quantity ?

2. What terrible consequences does it sometimes lead to ?

3. What bad permanent condition of the mind often occurs from alcohol ?

4. What peculiar tendency of mind does alcohol produce in families?

5. What number of persons per cent. in an asylum for the insane have been known to come there insane from partaking of strong drink?

6. What particular form of mental disease due to alcohol is commonly seen in the asylums?

SUMMARY OF LESSONS.

Now that we have learned so much about alcohol as it appears under the many disguises of strong drinks, we are, I trust, armed by our knowledge against its evil influences. We shall, however, still find many to defend the use of alcohol, for many, very many, are still ignorant about it; many, very many, are strongly prejudiced in favor of it; many, very many, are so fond of it they cannot help praising it as a good thing for themselves, and therefore as a good thing for everybody. Such is the strange perversity of the human mind, that numbers of people who are going wrong and who know they are going wrong, in the use of alcohol will still persist in their error, and with their eyes open to the wrong they are doing will persist in leading others with them. It is one part of the madness inflicted by alcohol on its friends, that it deceives them and in turn makes them deceivers.

You will have often in your lives to listen to the arguments of these persons. They will tell you a great deal of error, which you must be ready to hear, and at once recognize as error. You will be told that alcohol is a food because it warms the body. You know what that is worth. You know that alcohol only makes the body *feel* warm because it causes more warm blood to come to the surface of the body, there to lose its heat and leave the body colder. You know that cold and alcohol exercise the same kind of influence on the body, and that when working in the cold, even in the extremest cold, that man will work longest and best who avoids alcohol altogether.

You will be told that alcohol is a food because it gives strength to the body and helps men and women to do more work. You know what that is worth. You know that the

action of alcohol is to lessen the muscular power; that it
weakens the muscles, and that carried a little too far it dis-
ables them for work altogether, so that they cannot support
the weight of the body.   You know also from the experience
of men who have performed great feats of strength and en-
durance that such men have been obliged to abstain from
alcohol completely in order to succeed in their efforts, and
have beaten other men by reason of their careful abstinence.

You will be told that alcohol is a food because it makes
the body fat and plump and well nourished.   You know what
that is worth.   You know that there is nothing in alcohol that
can make any vital structure of the body; you know that the
best that can be said about alcohol in this matter is that in
some forms in which it is taken, as beer, for instance, it may,
because of the sugar in such drink, add fat to the body; and
you know that this is really not a good addition, because much
fat interferes with the motion of the vital organs, makes the
body heavy and unwieldy, and getting into the structure of
organs such as the heart or kidneys, makes those organs in-
capable of work, and so destroys life.

You will be told that alcohol makes you digest your food,
and helps people with weak digestion to enjoy their food and
digest it.   You know what that is worth.   You know that
every other animal except man can enjoy and digest food
without alcohol, and that men who never touch alcohol may
have excellent digestive power.   You know also that alcohol
impairs digestion, that in thousands of people it keeps up a
continual state of indigestion, and that the indigestion itself
is a temptation to these to take alcohol to a fatal excess.

You will be told that if alcohol be not a food in the strict
sense of the word, it is, notwithstanding, a luxury which a
man cannot do without with comfort to himself; that it cheers
the heart and is necessary for mirth and pleasure.   You know
what that is worth.   You know that young people like yourselves
can laugh and play and be as happy as the day is long without
ever tasting a drop of alcohol.   You know that hundreds of
men and women are as happy as they can be without a drop
of alcohol, and are much freer from worry, and anger, and
care about trifles than are those who take alcohol.   You know,
moreover, that after men or women have been cheered, as
they call it, by alcohol, they suffer a corresponding depres-

sion, and are made often so miserable that life is a burden to them until once again they have recourse to their cause of *short happiness and long sorrow.*

Lastly, whatever argument you may hear in favor of alcohol, you are now fully aware of its fatal power ; how it kills men and women wholesale, sending some to the grave straightway, and some to the grave through that living grave the asylum for the insane.

This is your knowledge. I would not advise you, as juniors, to intrude it in argument on your seniors, for that were presumptuous. But treasure it in your hearts. Let it keep you in the path of perfect abstinence from alcohol in every disguise, and believe me that your example will be all the more effective with older persons because it is a young example. Believe, finally, that you yourselves will, under the rule of total abstinence, grow up strengthened in wisdom, industry, and happiness, and that your success in life will reward you a thousand-fold for every sacrifice of false indulgence in that great curse of mankind—strong drink.

THE USE OF TOBACCO.

THE well-known article called tobacco is manufactured from a plant of the same name, which grows in warm climates, the largest production being in the Southern States of America and the West India Islands. The leaves of this plant, as chemically analyzed, yield nicotine, a colorless and intensely poisonous liquid, albumen, gum, resin, malic and citric acids, potash, lime, silica, and other constituents; and the smoke occasioned by burning these leaves contains, besides watery vapor, a certain quantity of free carbon, (very injurious to the throat and bronchial tubes,) ammonia, carbonic acid, and oil of tobacco, which latter is poisonous. Dr. Richardson declares that while there are no grounds for believing that the smoking of tobacco can produce any organic changes, it can and does produce various functional disturbances, in the stomach, the heart, the organs of the senses, the brain, the nerves, the mucous membrane of the mouth, (causing what has been described as the "smoker's sore throat,") and on the bronchial surface of the lungs.

In Taylor's "Principles and Practice of Medical Jurisprudence" it is declared that the effects which tobacco produces in large doses, or when taken by persons unaccustomed to its use, in the form of powder, infusion, or excessive smoking, are "faintness, nausea, vomiting, giddiness, delirium, loss of power of the limbs, general relaxation of the muscular system, trembling, complete prostration of strength, coldness of the surface, with cold, clammy perspiration, convulsive movements, paralysis, and death. In some cases there is purging, with violent pain in the abdomen; in others, there is rather a sinking or depression in the region of the heart, creating a sense of impending dissolution. With the above-mentioned symptoms there is a dilatation of the pupils of the eyes, dimness of the sight, a small, weak, and scarcely perceptible pulse, and difficulty of breathing."

The writer in the "American Cyclopedia" says: "The medical effects of tobacco upon the system are very marked, whether it is taken internally or applied externally. In small quantities, taken by either of the methods by which it is commonly used, as smoking, chewing, or snuffing the pulverized dry leaf, it acts as a sedative narcotic. In large quantities, or with those unaccustomed to it, the effect is giddiness, faintness, nausea, vomiting and purging, with great debility. As the nausea continues, with severe retching, the muscles relax, the skin becomes cold and clammy, the pulse feeble, and fainting and sometimes convulsions ensue, terminating in death. Its power in causing relaxation of the muscular system is great."

The "Manual of Hygiene," prepared by the Provincial Board of Health, and authorized for the use of schools and colleges in Ontario, says :—"Tobacco is injurious when used by young persons whose physical development is not completed. In persons unaccustomed to its use, increased flow of saliva, nausea, and muscular weakness are produced. When larger quantities are used, vertigo, general weakness, universal relaxation, depression, and increased frequency of the pulse, coolness of the surface, faintness, and vomiting ensue." * * * " Acute poisoning may result from the use of a large quantity, and sometimes death occurs. Dangerous and even fatal effects have resulted from the external application of fresh tobacco juice to the scalp, in cases of ring-worm."

This is medical testimony ; and it is borne out in the experience of thousands who use tobacco for the first time, or who use too much at any time. It may be objected that millions learn to like this article, and never feel such effects after the first few times of using, unless they take too much at once. That may be quite correct. And yet, is there not something very suggestive as to the character of an article which at first causes such symptoms? It is certain that the first using of milk, or meat, or fruit, does not create such distress. And it is a fair question whether a substance which creates such a profound sensation, antagonistic to health and comfort, when *first* taken into the system, is a proper article to continue taking, even though the unpleasant symptoms pass away with frequent use. This ought to be a strong argument against contracting this habit, even outside of the medical testimony to the numerous cases of dyspepsia, nervous disorders, palpitation, paralysis, cancer, and other afflictions, which physicians have traced directly to the use of tobacco.

But even if no evil effects of this description could be so traced to it, the utter *needlessness* of the habit, and its power-lessness to improve the character or promote the welfare of the person following it, should suggest to the intelligent young man that it is not wise or desirable to begin. He need not criticise older persons, who have contracted the habit; but should take care not to begin for himself. Many a man who is now a slave to the use of tobacco would give him the same advice. For it is a habit which *grows ;* and constant indul-gence renders the person powerless to resist the desire. A confirmed smoker is, moreover, very often disagreeable to those who do not use or like tobacco. By the odor of his breath, or by the fumes of his pipe as he passes on crowded sidewalks or other thoroughfares, he is very apt to cause offence to many, to whom tobacco in all its forms is disagreeable. Add to all this the expense. Even at ten cents a day,— a very moderate outlay,—it means $36.50 a year; and this is the interest at 6 per cent. on a capital of $600—too much money, the sensible young man will say, to sink in a practice which is absolutely needless, which is dis-agreeable to learn, which renders the person indulging it also disagreeable to very many of his fellow-beings, and which is often the cause of losses by fire, of slavery to a useless habit, of serious disorders, and, as the medical testimony proves, sometimes of insanity or death.

Questions on Lesson XLIII.

1. From what plant is tobacco manufactured, and where is this plant mostly cultivated ?

2. What constituents are found in the leaves of the tobacco plant ?

3. What two constituents, one found in the leaf, and the other in the smoke caused by burning it, are poisonous ?

4. What symptoms are produced in the person using tobacco for the first time, or in one using it excessively at any time ?

5. What inference may be drawn as to the character of an article which creates such disorders when first taken into the system?

6. What may be said as to the necessity of the habit of tobacco using, and of the acquisition of force and control by the habit itself?

7. Give some personal reasons, as furnished by the experience of tobacco users, against the contracting of the habit.

APPENDIX.

APPENDIX.

LESSON XLIV.

(Lesson XVI. in English Edition.)

COMPOSITION OF ALCOHOL.

THE chemists, after they had separated alcohol in the pure state, formed an idea that it was composed of water combined with what they called phlogiston or elementary fire. It was not unnatural that they should come to this conclusion, for when they set fire to alcohol, it burned away, giving out heat and flame, while all that was left seemed to them to be water. They did not know of the presence of the gas of which I spoke in the 15th lesson. To alcohol was, therefore, sometimes applied the term fire-water; and the term is a very expressive one, though, except by simile of expression, incorrect. We now know that alcohol burns in the air because it contains in itself two elements, that is to say two bodies, neither of which can by any known method be divided into anything further. Each of these elements is combustible, or capable of burning in the air.

The two elements are *carbon* and *hydrogen*. The element carbon in its separate state is a solid dark substance, like the soot which can be collected from the smoke of a burning candle, which is impure carbon. The element hydrogen in its separate state is a gas, the lightest of all known substances in nature. In alcohol these two elements are united together in the proportion of two ultimate or atomic parts of the carbon to five of the hydrogen. In this state of combination these elements constitute what is called a radical, by which is meant a substance that acts as if it were itself an element, though composed of two or more elements. The radical so formed has received the special name of *ethyl*.

In alcohol the radical ethyl is combined with the elements which constitute water in this manner. Water consists of two elements called oxygen and hydrogen, in the proportion of one ultimate or atomic part of oxygen to two of hydrogen. In their pure and distinct states both of these elements exist as gases ; but combined in the proportions described, they become a fluid, and exist in the fluid form we know so well as *water*. In alcohol, then, one of the atomic parts of the hydrogen of water is replaced by one part of the radical ethyl. So alcohol is after all a form of water : it is water changed in constitution or form by having in it carbon as well as hydrogen.

Water consists of three atomic parts, two of hydrogen and one of oxygen. Thus :

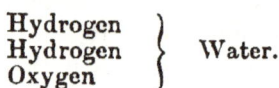

$$\left.\begin{array}{l}\text{Hydrogen}\\\text{Hydrogen}\\\text{Oxygen}\end{array}\right\} \text{Water.}$$

Alcohol consists of nine atomic parts—two of carbon, six of hydrogen, and one of oxygen. Thus :

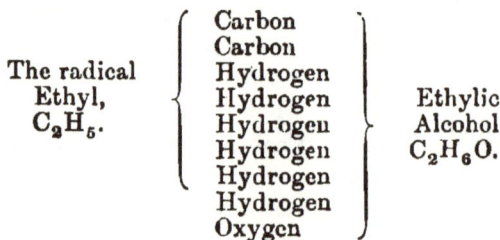

$$\text{The radical Ethyl, } C_2H_5. \left\{\begin{array}{l}\text{Carbon}\\\text{Carbon}\\\text{Hydrogen}\\\text{Hydrogen}\\\text{Hydrogen}\\\text{Hydrogen}\\\text{Hydrogen}\\\text{Hydrogen}\\\text{Oxygen}\end{array}\right\} \text{Ethylic Alcohol } C_2H_6O.$$

I hope this makes it clear to the mind in what respects alcohol resembles, and in what respects it differs from, water. Alcohol, it is seen, contains much more combustible matter than water, so it burns in the air. Alcohol is much less simple in its composition than water. Alcohol contains an element, carbon, which water does not contain. Alcohol contains more hydrogen than water.

The composition of alcohol may be changed by very simple means. If we take a little bit of the metal called sodium, or a little bit of the metal called potassium, and drop either of these bits of metal into pure alcohol, a gas escapes which burns. That gas is hydrogen. Meanwhile the metal itself dissolves and disappears, but it is not lost. It simply displaces the atom of hydrogen in the alcohol which was combined with

the oxygen. So now there is formed an alcohol, supposing sodium were
the metal added, of the following composition :

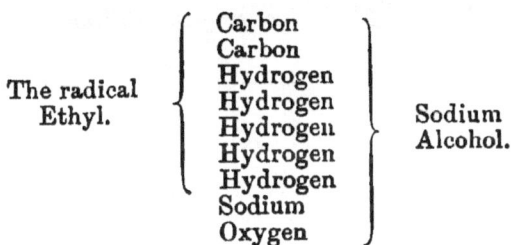

<pre>
                    ⎧ Carbon   ⎫
                    ⎪ Carbon   ⎪
     The radical   ⎪ Hydrogen ⎪
                    ⎨ Hydrogen ⎬   Sodium
       Ethyl.   ⎪ Hydrogen ⎪   Alcohol.
                    ⎪ Hydrogen ⎪
                    ⎩ Hydrogen ⎭
                      Sodium
                      Oxygen
</pre>

This is called sodium alcohol, but there is also a potassium alcohol,
and some other metallic alcohols, all constructed on a similar plan.

Questions on Lesson XLIV.

1. Of what did the early chemists consider spirits of wine or alcohol
to be composed ?

2. Of what do we now know alcohol to be composed, that is to say,
what are its elementary parts ?

3. State the meaning chemically of an element and a radical : in
what does a radical resemble, and in what does it differ from, an element ?

4. State the difference of composition between water and alcohol.

5. By what experiment can one part of the hydrogen of alcohol by
removed ?

6. What is the composition of the new substance so produced ?

LESSON XLV.

(Lesson XVII. in English Edition.)

PROPERTIES OF ALCOHOL.

As stated already, pure alcohol is a transparent fluid. It is of rather pungent odor, it is hot or burning to the tongue, and it burns in the air. It has still some other properties, which require to be learned. Compared with water, in respect to weight, alcohol is much the lighter fluid of the two. This fact is provable by an exceedingly simple experiment. Take a bottle that will just hold one thousand grains of water. Put that bottle in a fine balance, and poise; that is to say put as much weight in the opposite scale as will precisely balance the bottle. Then fill the bottle with water, in order to make sure that the water amounts to exactly one thousand grains. Next empty the bottle, dry it thoroughly, rinse it two or three times with a little pure alcohol, and dry it again. At last fill the bottle with pure alcohol, and weigh once more, retaining the poise of the bottle in the opposite scale. It will now be discovered that the bottle filled with alcohol will not weigh a thousand grains, as it did when it was filled with water, but will weigh seven hundred and ninety-two grains, or two hundred and eight grains less than the water. In scientific language, the weight of a liquid when compared with the weight of water, is called the specific weight. The specific weight of water is taken as one thousand by a standard of weight in a given measureful, and the specific weight of other fluids is calculated by the weight of the same measureful. If the measureful of other liquid be heavier than the measureful of water, we say the specific weight of that liquid is above that of water ; if the measureful be lighter, we say the specific weight is below that of water, and we state the difference above or below by the difference of grains or degrees. A thousand-grain measureful of water holds only seven hundred and ninety-two grains of alcohol, so the specific weight of alcohol is recorded as being 792°. Alcohol brought to this weight is termed absolute alcohol ; that is to say, it

108

has been brought to such purity that no water is left in it. What is ordinarily sold as pure alcohol is not refined to such a degree, so it weighs heavier, and may have a specific gravity or weight of 830°. Such alcohol is sufficiently rectified for most purposes, and great labor is saved by not extracting water from it beyond this point.

Alcohol in its pure state is not only lighter than water, but it boils at a lower temperature. This can be proved by another simple experiment. Place over a lamp a glass or porcelain vessel containing distilled water, and heat the water until it boils. When the water boils, put into it a thermometer graduated for high temperatures. It will be found that the thermometer, if it be on Fahrenheit's scale, will declare a register of 212 degrees of heat. We say, therefore, that water boils at 212° F., and we call that degree the boiling-point of water. If we have less heat the water will not boil; if we add more heat, the temperature of the water will not rise higher in an open vessel. The boiling-point is fixed and definite.

Next, into another glass or porcelain vessel placed over a lamp, and holding water, put a wide-mouthed glass flask containing alcohol. Now place a thermometer in the water, and another thermometer in the alcohol, and apply heat to the water. Watch both thermometers. It will then be found that when the thermometer in each rises to 172°, the water will not be boiling, but the alcohol will commence to boil. Below 172° the alcohol will not boil, and above 172° it will not increase in temperature if the boiling be not interfered with, and the mouth of the flask be quite open. Alcohol boils, therefore, at forty degrees of heat lower than water, and its boiling point is described as 172° Fahr. If, instead of using a thermometer marked on Fahrenheit's scale, one marked on the Centigrade scale were used, the water would be found to boil at 100°, and the alcohol 22° lower. The boiling point of alcohol is said, therefore, to be 78° Centigrade.

Questions on Lesson XLV.

1. What is the difference of weight between alcohol and water?
2. How can this difference be proved?
3. What is meant by the specific weight of a fluid substance?
4. What is the heat at which water, and what is the heat at which alcohol boils?
5. What process will prove this?
6. What is meant by the term boiling-point? State the boiling-points of water and alcohol on the Fahrenheit and Centigrade scales.

LESSON XLVI.

(Lesson XVIII. in English Edition.)

ALCOHOLIC DRINKS.

THE alcohol we have studied so far is the alcohol derived originally from wine by the process of distillation, and is called ethyl or ethylic alcohol. It forms the active part of all our common drinks, and it may be obtained from any of them by distillation.

The amount of alcohol in these different drinks varies very much indeed, and so they vary in their power to produce intoxication or drunkenness.

We estimate the amount of alcohol in wine, beer, or spirits by one of two standards, by weight or by volume per cent. If we estimate by weight, we find how many grains of absolute alcohol, that is alcohol freed of the last traces of water, there are in one hundred grains of the wine, spirit, beer, or other drink under observation; and when we have determined that fact, we say that the percentage of alcohol in the specimen before us is so much by weight. From a specimen of what is called British ginger wine I obtain ten grains of alcohol from every hundred grains of wine; so I say of such wine that it is of ten per cent. strength by weight.

If we estimate by volume, we find what number of volumes of alcohol, each volume being a measure of one hundredth part, there are in a one hundred volume measure of the wine.

From the same specimen of British ginger wine above named I obtain the fact that in every hundred measures, there are thirteen measures and a half of absolute alcohol. So I say of such wine that it is of thirteen and a half per cent. strength by volume.

Some persons get confused when they hear of the differences of percentage by weight and by measure; but when it is remembered that if two fluids of different specific weights be weighed out in equal proportions, say a hundred grains of each, the lightest fluid will fill the largest space, there need be no difficulty.

110

It is best, I think, to remember the strength of different alcoholic liquids by the volume rather than by the weight, because when persons take intoxicating drinks they do not, as a rule, weigh them. I am sorry to say they do not measure them with much accuracy ; for what they do in this way is usually by measure. "A tumbler of ale, a glass of wine, a gill of whiskey, a quartern of rum "—this is the mode in which strong drinks are called for by those who indulge in them and spend their money on them. It will therefore be as well to keep in mind the measure, and to state the amount of alcohol in different drinks by the volume of it rather than by the weight.

The alcohol in intoxicating drinks is, as we now know, the cause of intoxication ; but it will be asked, what, then, is the rest of the liquid composed of ? This has been pretty clearly stated already. The rest of the liquid, in a large number of instances, is practically nothing but water. But some very luscious wines contain a large percentage of sugar, and almost all alcoholic drinks contain more or less of sugar.

Madeira contains five per cent. of sugar ; sherry, three per cent. ; port, three and even four per cent. ; some champagnes, ten per cent.

Certain wines contain earthy substances in a small proportion, and a few have traces of the metal iron.

Every kind of alcoholic drink contains a small quantity of free acid. Beers and ales contain free acid, and sugar and extractive substances.

These are natural parts of the drinks in question, and in the strong and refined wines and spirits there are volatile ethers. But, as a whole, all strong drinks produce their peculiar effect by reason of the alcohol they contain in combination with the water of which their bulk is made up. They act practically as mixtures of ethylic alcohol and water.

Questions on Lesson XLVI.

1. What is the scientific name given to the alcohol which enters into the common intoxicating drinks ?

2. By what standards is the amount of alcohol stated in different drinks containing alcohol ?

3. What is the difference between percentage by weight and by volume ?

4. How are alcoholic drinks usually divided out for sale ?

5. With what fluid is alcohol mainly combined in intoxicating drinks ?

6. What other substances are found in such drinks ?

LESSON XLVII.

(*Lesson XXII. in English Edition.*)

THE ALCOHOL FAMILY.

So far we have considered alcohol as if it were a single, simple fluid, and, for our purposes, it has been quite right to speak of it in this manner, because we have dealt with the alcohol first discovered by man in wine, the alcohol that is most largely used, and is still most generally known. We have dealt with the spirit or alcohol of wine, or with what the chemist calls "ethylic alcohol." The time is come when it is proper to state that there are many alcohols, and that by the light of modern science we are now made familiar with a whole family of chemical substances to which the word alcohol is applied.

All the alcohols, all the members of the alcohol family, are constructed on the same plan or type, and are got by fermentation from organic substances. The one alcohol with which we have become conversant thus far is obtained by the fermentation of sugar, of fruit, and of grain. The rest are obtained from other organic bodies.

METHYLIC ALCOHOL.

To begin with the first of the members of the alcohol family, we must go to what is called "wood spirit," or "pyroxylic spirit," or, as it is chemically named, "methylic alcohol." This spirit is obtained by the distillation of wood. It has not been known for more than sixty-five years. It was discovered by Mr. Philip Taylor in the year 1812. It boils at a lower temperature than ethylic alcohol, viz., at 140° Fahr. It has an aromatic smell, and it is slightly acid. It burns as the better-known alcohol burns, without giving off any smoke when it is quite pure; and, mixed with ordinary spirit, it is much employed for spirit-lamps, and for various purposes in the arts. It is sold under the name of methylated spirit, or as a part of the fluid that goes under that name.

We have seen that ethylic alcohol is composed, chemically, of a radical of carbon and hydrogen, known as ethyl, combined with oxygen. Methylic alcohol is, in like manner, composed, chemically, of a radical of carbon and hydrogen, combined with oxygen. But this radical is made up of one part of carbon and three of hydrogen, and is called methyl.

Methylic alcohol is composed as follows :

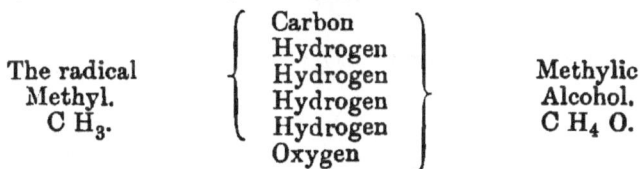

The radical Methyl. $C H_3$.	$\left\{\begin{array}{l} \text{Carbon} \\ \text{Hydrogen} \\ \text{Hydrogen} \\ \text{Hydrogen} \\ \text{Hydrogen} \\ \text{Oxygen} \end{array}\right.$	Methylic Alcohol. $C H_4 O$.

Methylic alcohol is not very pleasant to the taste, but it has, nevertheless, sometimes been taken as a drink by persons of singular taste. A friend of mine had a patient who indulged in methylic alcohol every day, preferring it to the ordinary alcohol of wine. I found, by testing the action of this alcohol, that although it is very injurious, it is not so injurious to the living body, when it is quite pure, as ethylic alcohol is. When pure methylic alcohol is placed on the lips, it produces a sensation of burning, in the same way as ethylic alcohol does, but not in the same degree. The taste of it is something like the taste of whiskey diluted with a fourth part of water. It has what is commonly called a smoky taste. Taken in free quantities, it produces intoxication, and its action in this respect is very rapid, because, owing to its physical properties, it diffuses freely. For the same reason, the intoxication caused by it is, comparatively, of short duration.

Like ethylic alcohol, methylic is soluble in water in all proportions, and it mixes with water with extreme readiness.

Questions on Lesson XLVII.

1. What has modern science taught us as to the extension of the use of the word alcohol?

2. The name of the first alcohol in the family of alcohols is methylic alcohol. By what means is methylic alcohol obtained?

3. In what respects does methylic alcohol differ from ethylic?

4. What is the elementary composition of methylic alcohol?

5. What is the taste and smell of pure methylic alcohol?

6. What are the general uses to which methylic alcohol is applied? What are its effects on the body when it is taken in free quantities?

LESSON XLVIII.

(*Lesson XXIII. in English Edition.*)

THE ALCOHOL FAMILY.—Continued.

THE alcohol particularly dwelt upon in the last lesson, and which is called methylic alcohol, was seen to be a lighter spirit than the common alcohol got by the distillation of wine. It contains, as we learned, one part of carbon, four of hydrogen, and one of oxygen, while ethylic alcohol contains two parts of carbon with six of hydrogen and one of oxygen. We have now to consider some other alcohols which are obtained by the distillation of another class of organic substances. These are made up of the same elements, but in different proportions; and by virtue of this change of construction, they are heavier fluids than those that have been described. They are also different in some other particulars of a physical character.

When molasses, grain, or other fermentable substances are subjected to the process of fermentation there may be produced several bodies of the alcohol series, each one of them distinguished from the rest by its weight, boiling point, and other physical qualities. Generally speaking, the first product of the distillation is called crude spirit. When this crude spirit is put into a retort, and is subjected to the process of redistillation, different fluids, different alcohols, distil over according to the degree of heat that is employed in the distillation, and so the crude liquor is divided into different parts which may be separately collected by repeated distillations.

It may be well to tell those who have not yet learned anything about chemical work, that this method of dividing one liquor, which at first appears to consist of one part only, into several parts, by distillation, is called the process of fractional distillation, a very simple and expressive name for a very refined and ingenious device. Whenever two or more liquids, possessing different weights and boiling-points,

114

will mix together to make what appears to be a simple fluid, they admit, as a rule, of being broken up by this plan of fractional distillation. The different fluids that are commingled pass into vapor at different degrees of heat, and are separable on subjection to the fractional distillation.

When, then, crude spirit is distilled by the fractional plan there are passed over, so as to admit of being collected in passing, different alcohols. In distilling French brandy the main product obtained is the liquid named brandy, *eau de vie*, which liquid contains over fifty per cent. of ethylic alcohol. But when the distillation is continued there is obtained another alcohol as well as the ethylic, which is heavier than the ethylic, of more pungent odor, and of a more oily and disagreeable taste. This alcohol is called, chemically, "propylic alcohol." It is composed of a radical of carbon and hydrogen, named propyl, combined as before with oxygen. There are in this radical three parts of carbon and seven of hydrogen. The composition of propylic alcohol may therefore be described as follows :

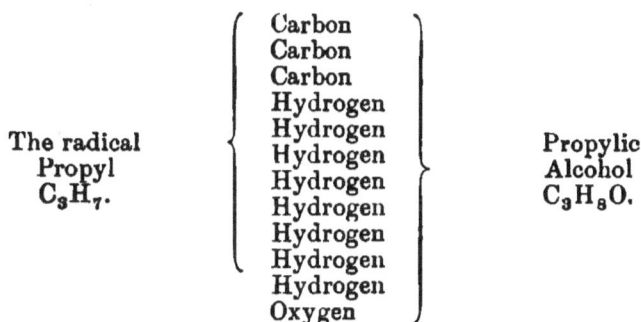

The radical Propyl C_3H_7.	Carbon Carbon Carbon Hydrogen Hydrogen Hydrogen Hydrogen Hydrogen Hydrogen Hydrogen Oxygen	Propylic Alcohol C_3H_8O.

Propylic alcohol is less soluble in water than the methylic and ethylic alcohols are. It burns with a little smoke, boils at 195° Fahr., and has a specific weight of 758, compared with water as 1000. When it is taken into the living body freely, it intoxicates as ethylic alcohol does, but the effects of it are more severe and prolonged, because it does not escape from the body so readily.

Questions on Lesson XLVIII.

1. What is meant by the process of fractional distillation ?
2. To what purpose is the process applied in respect to the alcohols?
3. What is the composition of the first of those alcohols which are heavier than the ethylic, viz., propylic alcohol ?

4. How is propylic alcohol obtained?

5. What are the physical properties of propylic alcohol as compared with ethylic?

6. What is the action of propylic alcohol on the living body?

LESSON XLIX.

(Lesson XXIV. in English Edition.)

THE ALCOHOL FAMILY.—Continued.

WE have not yet quite finished with the alcohols as a family. There are two more members of it, at least, with which we must become acquainted.

When the crude spirit derived from the fermentation of beetroot treacle is distilled, there passes over in vapor, at the latter part of the distillation, a heavy alcohol, which is called butylic alcohol. This alcohol is composed of a radical of carbon and hydrogen known as butyl, combined with hydrogen and oxygen. But the radical butyl contains four parts of carbon and nine of hydrogen, so that the composition of butylic alcohol is as follows:

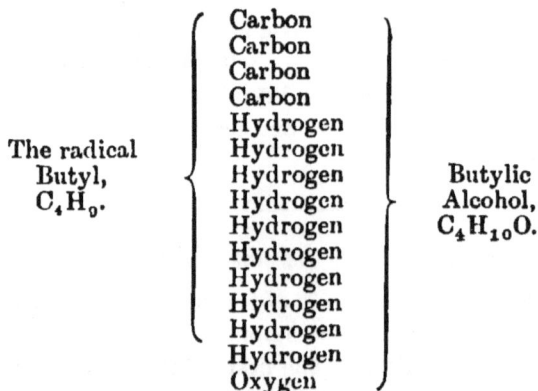

The radical
Butyl,
C_4H_9.

$\left\{\begin{array}{l} \text{Carbon} \\ \text{Carbon} \\ \text{Carbon} \\ \text{Carbon} \\ \text{Hydrogen} \\ \text{Hydrogen} \\ \text{Hydrogen} \\ \text{Hydrogen} \\ \text{Hydrogen} \\ \text{Hydrogen} \\ \text{Hydrogen} \\ \text{Hydrogen} \\ \text{Hydrogen} \\ \text{Hydrogen} \\ \text{Oxygen} \end{array}\right.$

Butylic
Alcohol,
$C_4H_{10}O$.

Butylic alcohol has a pungent aromatic odor and a hot biting taste. It burns, yielding a smoke, and if the smoke be received on a cold plate it gives a deposit of soot which consists of fine carbon. It has a specific weight of 803, compared with water as 1000, and it boils at 230° Fahr.; it is very insoluble in water.

This alcohol, when it is taken into the living body, produces a peculiar intoxication, attended with severe tremblings of the muscles. I makes the body exceedingly cold, and recovery from the bad effects of it is very slow.

By the fermentation of potatoes, as well as by the fermentation of beet, molasses, and of grapes, there is produced a fluid substance which remains when the lighter alcohols have been distilled over, and to which the name of crude potato spirit or fusel-oil has been applied. This fluid is of oily appearance, and is of a most nauseous odor. The smell of it adheres to everything that is touched by it for a long time. The crude fluid is made up of several parts, but it consists chiefly of a distinct alcohol which can be separated by careful distillation, and which is called " amylic alcohol."

Amylic alcohol is composed of a radical of carbon and hydrogen known by the name of amyl. This radical consists of five parts of carbon and eleven of hydrogen, and is combined in the alcohol, as in all previous instances, with one part of hydrogen and oxygen. The composition of this alcohol may therefore be expressed in the following manner :

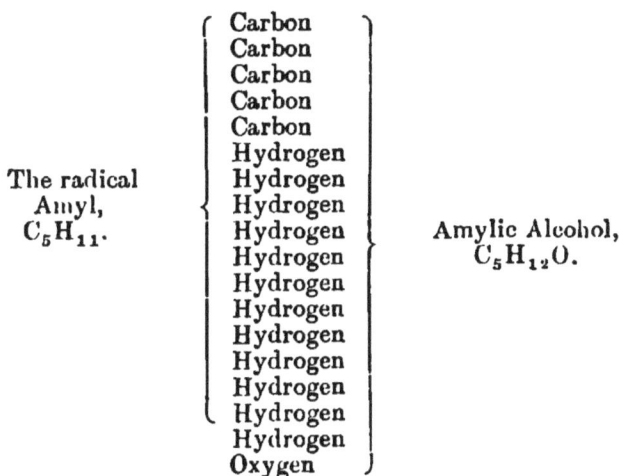

The radical Amyl, C_5H_{11}.

$$\left\{ \begin{array}{l} \text{Carbon} \\ \text{Carbon} \\ \text{Carbon} \\ \text{Carbon} \\ \text{Carbon} \\ \text{Hydrogen} \\ \text{Hydrogen} \\ \text{Hydrogen} \\ \text{Hydrogen} \\ \text{Hydrogen} \\ \text{Hydrogen} \\ \text{Hydrogen} \\ \text{Hydrogen} \\ \text{Hydrogen} \\ \text{Hydrogen} \\ \text{Hydrogen} \\ \text{Oxygen} \end{array} \right\}$$

Amylic Alcohol, $C_5H_{12}O$.

Amylic alcohol, when pure, is quite transparent, of pungent odor, and burning nauseous taste. When kindled into flame it yields an

abundant smoke and copious deposit of soot—carbon.   It has a specific
weight of 811, taking water as 1000, and it does not boil until it is exposed to a heat of 270° Fahr.   It is insoluble in water.   Amylic alcohol,
when taken into the body, causes an intoxication of a most dangerous
kind.   The muscles of the body are made tremulous, and the whole
body is rendered deathly cold.

In bad specimens of common spirits, such as whiskey, this alcohol is
present, and it renders them very hurtful and dangerous.

Questions on Lesson XLIX.

1. What is butylic alcohol, and how is it prepared ?
2. What are the chemical and physical properties of butylic alcohol?
3. What is the action of butylic alcohol on the living body ?
4. What is amylic alcohol, and how is it prepared ?
5. What are the chemical and physical characters of amylic alcohol?
6. What is the action of amylic alcohol on the living body, and in
what common spirituous drinks is it sometimes present ?

LESSON L.

(*Lesson XXV. in English Edition.*)

CONSTRUCTION OF THE ALCOHOLS.

THERE are a great many more alcohols in the alcohol family beyond those I have named. They keep increasing in weight in regular proportion of increase of the parts of carbon and hydrogen, until at last they become actually solid bodies, like wax. There is an alcohol to be got from spermaceti, called cetyl alcohol ; it is solid, white, and crystalline. To show how the carbon and hydrogen increase in their atomic parts in these alcohols, I may state that the last-named member of the group, cetyl alcohol, is composed of a radical containing sixteen atoms of carbon and thirty-four of hydrogen, combined, as before, with one atom of hydrogen and one of oxygen.

Whether the heaviest of the heavy alcohols will produce intoxication when they are introduced into the living body I cannot tell, because the experiment as to their action has not been made. I have followed up my researches with some of the alcohols that are heavier than amylic, and have found them all poisonous intoxicants, as far as I have been able to trace them ; but the opportunity has not been afforded me to try beyond two of the series above the amylic alcohol.

It will not be necessary, therefore, to say more as to the action of the alcohol family, beyond that member of it called amylic alcohol. The rest of the members are out of the way of every-day experience, and, for the matter of that, out of the way of ordinary skilled experience. I may leave them, and finish this account of the alcohols by pointing out the reasons why they are grouped together as a family.

They are grouped as a family, because they have certain similarities which bind them together in one bond. They are all, it will be observed, made or constructed on the same plan. A radical composed of carbon and hydrogen, in every instance, takes the place of one of the atoms of hydrogen of water. Further, it will be seen that each different radical of each series differs only in one particular from the others, namely, that the parts or atoms of carbon and hydrogen keep increasing, from the lowest up to the highest, in a regular order of progression. In every case the carbon increases by one step, the hydrogen by two steps, throughout the series.

All the alcohols burn in the air, but some of them burn without and some with smoke. The reason of this is very simple and interesting. In the process of burning, the carbon combines with the oxygen of the air, and the heat and flame are the consequences of that combination. If, in burning, the quantity of oxygen that reaches the carbon is sufficient to combine with all the carbon, there is complete combustion, and no production of smoke. In the case of the lighter alcohols, the methylic and ethylic, the quantity of carbon is so small that the combustion is complete. In the heavier alcohols, such as the amylic, the quantity of carbon present is so great that the combustion of the carbon is not complete ; the chemical form of the alcohol is broken up, and a part of the carbon is oxydized, but another part of the carbon is simply separated from its hydrogen without being consumed, and so it separates in the form of smoke or soot ; that is to say, it separates as carbon in nearly a pure state.

Once more, all these alcohols, when diluted with water and exposed, under favoring conditions, to the oxygen of the air, become acids. We say of alcohol of wine, when it becomes sour, that it has turned into vinegar, or acetic acid. Every other alcohol goes in like manner into an acid or vinegar of its own, under proper conditions for the change, and every alcohol has its own acid or vinegar. These likenesses are sufficient to mark off the alcohols as a family group.

Questions on Lesson L.

1. Give the name of a solid alcohol, and state from what it may be obtained.

2. Why do some alcohols become heavier and denser than others?

3. In cetyl alcohol, how many parts or atoms of carbon and hydrogen are combined together to make its radical?

4. State the reasons why the different alcohols are classed together in one family group.

5. Why do some alcohols burn without, and others with, production of smoke and soot?

6. What occurs to all the alcohols when they are diluted with water and exposed to the oxygen of the air under conditions favorable to change of composition?

www.ingramcontent.com/pod-product-compliance
Lightning Source LLC
Chambersburg PA
CBHW022103210326
41518CB00039B/710